Sustainable Development through Innovative Agriculture

The Editors

Dr C. Karthikeyan completed his M.Sc. (Ag.) and Ph.D., in Agricultural Extension at Tamil Nadu Agricultural University (TNAU), Coimbatore. Won Post Doc Fellowship to work on a project at Rockefeller Centre, Italy. Completed Australian Government's Competitive Post Doc. Fellow at Charles Sturt University, Australia. Graduated MBA (HRM) at Bharathiar University, Coimbatore. He is certified by AIIMA, New Delhi and recognised Fellow of the Indian Society of Extension Education, New Delhi.

Dr Karthikeyan currently, working as Professor and Head, Department of Social Sciences, at Agricultural College & Research Institute, TNAU, Thoothukudi, Tamil Nadu. Through World Bank support he pilot tested the concept 'e-Velanmai' (e-Agriculture) upscaled in 100 blocks by the Department of Agriculture, Tamil Nadu. Won seven International awards and 17 National/State level awards. He had implemented 19 externally funded researches and visited to countries in six continents of the world. So far published 16 ISBN books, 14 book chapters, 16 technical books, 104 Scientific and popular articles in International and National journals of repute.

Dr R. Sendilkumar completed his M.Sc. (Ag) and PhD, in Agricultural Extension from Tamil Nadu Agricultural University, Coimbatore. He also obtained MBA and PGDIM from Bharathiar University, Coimbatore and PGDPM from Annamalai University, Chidambaram. He was an ICAR Junior fellowship holder during 1991. He is an active life member and Office bearer in various reputed professional bodies. 18 years of experience in Teaching, Research, Extension, Farm management and Administration. Recognized faculty in Agribusiness, guided MSc, MBA and PhD students. Presently working as Professor Agrl. Extension at Kerala Agricultural University. Published one monograph, co-authored two books, 10 book chapters and 41 research papers in reputed National and International journals. Specialised in the area of Agricultural risk management education, Value chain management, Agribusiness and Entrepreneurship, Research Methodology and Group Dynamics.

Sustainable Development through Innovative Agriculture

– Editors –

Dr C. Karthikeyan

Professor & Head
Department of Social Sciences,
Agricultural College & Research Institute,
TNAU, Thoothukudi,
Tamil Nadu

Dr R. Sendilkumar

Professor (Agrl. Extension),
College of Cooperation, Banking and Management,
Kerala Agricultural University, Vellanikkara ,
Thrissur, Kerala

2019

Daya Publishing House®

A Division of

Astral International Pvt. Ltd.

New Delhi – 110 002

ISBN: 9789388173872 (Int. Edition)

Publisher's Note:

Every possible effort has been made to ensure that the information contained in this book is accurate at the time of going to press, and the publisher and author cannot accept responsibility for any errors or omissions, however caused. No responsibility for loss or damage occasioned to any person acting, or refraining from action, as a result of the material in this publication can be accepted by the editor, the publisher or the author. The Publisher is not associated with any product or vendor mentioned in the book. The contents of this work are intended to further general scientific research, understanding and discussion only. Readers should consult with a specialist where appropriate.

Every effort has been made to trace the owners of copyright material used in this book, if any. The author and the publisher will be grateful for any omission brought to their notice for acknowledgement in the future editions of the book.

Published by : **Daya Publishing House®**
A Division of
Astral International Pvt. Ltd.
– ISO 9001:2015 Certified Company –
4736/23, Ansari Road, Darya Ganj
New Delhi-110 002
Ph. 011-43549197, 23278134
E-mail: info@astralint.com
Website: www.astralint.com

Digitally Printed at : **Replika Press Pvt. Ltd.**

Sustainable Development through Innovative Agriculture

– *Editors* –

Dr C. Karthikeyan

Professor & Head
Department of Social Sciences,
Agricultural College & Research Institute,
TNAU, Thoothukudi,
Tamil Nadu

Dr R. Sendilkumar

Professor (Agrl. Extension),
College of Cooperation, Banking and Management,
Kerala Agricultural University, Vellanikkara ,
Thrissur, Kerala

2019

Daya Publishing House®

A Division of

Astral International Pvt. Ltd.

New Delhi – 110 002

ISBN: 9789388173872 (Int. Edition)

Publisher's Note:

Every possible effort has been made to ensure that the information contained in this book is accurate at the time of going to press, and the publisher and author cannot accept responsibility for any errors or omissions, however caused. No responsibility for loss or damage occasioned to any person acting, or refraining from action, as a result of the material in this publication can be accepted by the editor, the publisher or the author. The Publisher is not associated with any product or vendor mentioned in the book. The contents of this work are intended to further general scientific research, understanding and discussion only. Readers should consult with a specialist where appropriate.

Every effort has been made to trace the owners of copyright material used in this book, if any. The author and the publisher will be grateful for any omission brought to their notice for acknowledgement in the future editions of the book.

Published by : **Daya Publishing House®**
A Division of
Astral International Pvt. Ltd.
– ISO 9001:2015 Certified Company –
4736/23, Ansari Road, Darya Ganj
New Delhi-110 002
Ph. 011-43549197, 23278134
E-mail: info@astralint.com
Website: www.astralint.com

Digitally Printed at : **Replika Press Pvt. Ltd.**

Professor K.V. Peter
TC-12/1309 Mullakkara P.O. Mannuthy Thrissur – 680 651
Former Professor of Horticulture and Former Vice-Chancellor
Ph.D., FNAAS, FNASc, FNABS, FISVeg, FISGen., FHSI

KERALA AGRICULTURAL UNIVERSITY
Main Campus, Kerala Agril. University
P.O. Thrissur - 680 656, Kerala

Phone: (R) 0487-2373017
E-mail: kvptr@yahoo.com;
peter.kv@gmail.com
www.divinenoni.com

Foreword

Uncertain rainfall and consequent drought in the last three years (2014-17) have made farming more risky and challenging than ever before. Recent statistics show that the farmers quitting agriculture and moving to cities is a whopping and disturbing 62 per cent. The growing distress and declining confidence among these small and landless farmers are due to low returns on investments and farmer unfriendly marketing. The producer farmer has no control on retail prices, the bulk pocketed by middlemen. Adding to this, there is over dependency on nature for soil, water and energy .The irrigation systems are ill maintained and wastage colossal though the theme of scientific water management is "one drop one crop". The current agricultural practices in rural areas are neither economically nor environmentally sustainable and hence India's yields for many agricultural and horticultural crops are low and economically unsustainable. Mere general fixes like subsidies, procurement policies, minimum support prices (MSP) and loan waivers have not served their purpose as the share of price paid by the consumers and received by the farmers are abysmally low. The issue in-fact questions the sustainability of Agriculture in the long term.

The overall agricultural sustainability indices during 2011 was seen high in Himachal Pradesh (0.64), followed by Kerala (0.61) and Punjab (0.55). Sustainability in agriculture is largely understood with direct or indirect contributions of different indicators. Despite several researchers drafting out different frame work to study the sustainability, consensus is yet to be emerged to have common indicators, because of different agro climatic zones, methods and systems of farming and extent of mechanization. It is strongly recognized that application of innovations-indigenous and scientific- in agriculture and extent of their economic and eco-friendly uses will set a road map for ensuring sustainability. Development of a need based innovation is the process of putting a new idea in a form expected to meet the requirements of farmers as potential adopters.

In general, consensus is emerging among the stakeholders-farmer producers, traders and consumers- and thrust is given to develop innovative technologies, which do not create adverse effects on environmental resources and accessible to all kinds of farmers - small and marginal farmers-. Improvising crop productivity to meet burgeoning population, ensuring food and nutrition security and providing availability, access and absorption (3 As) are the objectives of any innovations.

The DNA of innovative agriculture should be fabricated with three fold sustainability dimensions viz., ecological sustainability, economic sustainability and social sustainability. If an agricultural innovation developed with said elements in the DNA of technology generation, dissemination and utilization will pave way for sustainable agriculture.

This is the right time to mine out, whether we have adequate stock and documentation of innovative technologies to address the sustainability issues in agriculture and to reach the Sustainable Development Goals. I am happy to note that authors have rightly sensitized the need of the hour and brought out the publication entitled *"Sustainable Development through Innovative Agriculture"* Prof. (Dr) C. Karthikeyan, Professor of Agricultural Extension, Tamil Nadu Agricultural University and Prof. (Dr) R. Sendil Kumar, Professor of Agricultural Extension, Kerala Agricultural University are well known scientists and teachers who devoted considerable time to author chapters and edit the book. Forty nine working scientists from two Universities and 15 centres of knowledge contributed 23 chapters making the book authoritative, reader friendly and above all useful to students, policy planners and farmers. I compliment Prof. (Dr) C. Karthikeyan and Prof. (Dr) R. Sendil Kumar for the efforts made to bring out the present book.

I hope this compilation will form the stepping stone to reach the Sustainable Development Goals (SDGs) of our nation.

I take the opportunity to congratulate the publisher "Astral International Pvt Ltd" New Delhi known for meticulous and error free printing. I have decade long association with the publisher.

Prof. (Dr.) K.V. Peter

Former Vice-Chancellor
Kerala Agricultural University

Preface

UN General Assembly by the UN Open Working Group (OWG) 2015 has redefined and transforms the Millennium Development Goals (MDGs) in to Sustainable Development Goals (SDGs). Accordingly, 17 SDGs are available before the world for the future course of action and vigilance to achieve the same. All the goals prescribed are interconnected and interwoven in such a way that complementary and supplementary to each other. Especially the goals deals with the aspects like poverty, hunger and food and nutritional security, health, education, gender and empowerment, resource conservation and combating climate change *etc.*, requires the intervention of innovative technology in agriculture, which necessarily foster the sustainable agriculture development. To pace the development, technology should be innovative and sustainable in order to conserve the resources and not to deplete. Since resources are scarce and has to be conserve with the social responsibility motive to made available for the future generation. On that way technology creation, development and dissemination, innovative approaches are much needed. The learned experience over years on various issues viz., complex, diverse, and risks prone nature of agriculture, needs sustainable innovative technology to combat. Therefore social responsibility of Scientist urges to create and develop sustainable technology in agriculture and allied. Hence, an earnest attempt was taken to present the emerging promising innovative technology, institutional intervention in agriculture. This compilation will be greatly useful for the stake holders like technology producers, adopters, agricultural educationist, extension Professionals, students, research scholars, policy makers and administrators. In this occasion authors take the opportunity to acknowledge the contributors of the various chapters of this book and also the publisher Astral International Pvt Ltd., New Delhi.

Dr. C. Karthikeyan

Dr. R. Sendilkumar

Contents

Chapter 1

Innovative Extension Approach for Sustainable Livelihood of Farmers in Tamil Nadu

C. Karthikeyan

Professor (Agrl. Extension)
e-Extension Centre
Tamil Nadu Agricultural University
Coimbatore – 641 003, Tamil Nadu

Introduction

It is a matter of great concern that there exists wider gap between the proportion of the technologies generated by the public agricultural research system and technologies adopted by the farmers in India. Supporting this fact, the all India value of extension effectiveness index was 47 per cent indicating moderately effective extension mechanism in India (Mishra, 1996). One of the vital reasons for such wide gap is due to the lack of an appropriate extension/technology transfer model in Agriculture that could establish better linkage between Scientists and farmers in order to facilitate effective dissemination of location specific appropriate technologies in a timely manner. Hence there is a need to develop suitable approach for technology transfer in agriculture. Considering this research issue, an attempt was done to conceive an Information and Communication Technology (ICT) based extension model called 'e-Velanmai' (means e-Agriculture) for dissemination of farm specific agricultural technologies from the agricultural scientists to the needed farmers in the selected command areas of Tamil Nadu state. This model was subjected to pilot testing and validation to standardize the technology transfer process for achieving maximum effectiveness in solving the farm problems leading to enhanced returns to farmers from agriculture. The pilot experiment was done during July 2007 to March 2011 in three sub basins *viz.*, Palar, Varaghanadhi and Aliyar of Tamil Nadu with the support of the World bank aided TN-IAMWARM project of the Government of Tamil Nadu. Based on the successful results obtained in the performance of the model, World Bank has supported for upscaling the e-Velanmai model of extension in 19 irrigation project command areas of the state during 2011-12 and in 26 sub basins during 2012-13. 'e-Velanmai'project was evaluated periodically and recommended by World Bank experts for implementation as a technology transfer model during 2014-15 in 100 blocks of Tamil Nadu through the extension officials (BTM/ATM) of the DOA. This chapter presents about the description of the model and its performance in Tamil Nadu.

The specific objectives are:

- ✰ To upscale the 'e-Velanmai' model of technology transfer to farmers in selected command areas of Tamil Nadu and
- ✰ To document the farmer's perception about the ICT based technology transfer model.

Concept of e-Velanmai

It is a combination of personal and ICT based, demand driven, participatory and sustainable extension approach to provide appropriate and timely agro advisory services by scientists to the registered farmers using ICT tools (Internet, Tablet, Mobile Phone *etc.*) on need and/or regular basis with necessary follow up actions attempted by Field Extension Coordinators (Karthikeyan, 2012).

"e-Velanmai" is described as a combination of personalized and ICT based extension approach due to the presence of human element called Field Coordinator (FC) who handles ICT tools to link farmers and experts for the purpose of technology transfer.

It is also referred as demand driven, participatory extension approach because:

- ✰ Farmers paid an annual membership fee (₹50-150) based on the farm size owned by them to avail the extension services under 'e-Velanmai'. The nominal fee was collected to ensure the participation of farmers in the project.
- ✰ Scientists attended the farmers queries based on their call (demand) or need and hence it is demand driven for technical advice.

It is also believed to be sustainable approach of extension as:

- ✰ It is envisioned that the farmers could operate and follow the e-Velanmai model of technology transfer by them to access technologies in the long run. The membership fee collected could be utilized to achieve sustainability of the approach. Hence sustainability component is inbuilt in the model.

- Moreover action research on sustainability of the project was attempted and the results indicated that Public-Public model (TNAU and DOA, TN) was sustainable to render cost effective and quality extension service to the farming community.

Membership Fees

Farmers who were interested to avail the services of 'e-Velanmai' were supposed to pay an annual membership fee of ₹50 to ₹150 per farmer depending on the size of their farms (Table 1.1). Fee structure of ₹50-300 per farmer was decided by the farmers themselves through two brainstorming meetings held at village and WUA level held at Sencherimalai village and Kumarapalayam WUA respectively of Palar sub basin during September 2008. However it was revised as ₹50-150 during the conduct of the planning workshoponupscalingof e-Velanmai. Fee collection was done mainly to enhance the sustainable participation of farmers in the scientific farming through e-Velanmai model of extension. The fee structure for availing extension services was decided as follows:

Table 1.1: Membership Fee Structure in e-Velanmai

Farm Size (acres)	*Membership Fees ₹/Farmer/Year*
Upto 5	50
5-10	100
>10	150

Those farmers who were interested to avail the ICT based extension advices were considered as members and enrolled in the scheme. Wherever there existed a WUA, its farmers were gathered for an awareness meeting about the e-Velanmai and those interested farmers were enrolled as members in the scheme. Each member was given a membership card containing his farm details and a record of the date of visit of the FC, problems observed and technical advices given to him by the scientists. The membership money was deposited in the WUA account for ensuring sustainability of 'e-Velanmai' activities by the WUA beyond project period.

Implementation of the e-Velanmai Scheme (Pilot testing)

Cost Free Model

During July 2007 the scheme was implemented in Palar sub basin by selecting five villages spread over five water users association covering 25 farmers allotting five farmers from each village. One FC was appointed to offer technical advices from the experts to the farmers. The FC visited one village per day to offer extension advice to five farmers thus covering all the 25 farmers in a week. The expert team of Scientists comprising of Entomologist, Soil Scientist, Pathologist, Crop Physiologist, Agronomist and Horticulturist was set up at TNAU, Coimbatore to offer technical advice considering the damaged specimen photographs captured using digital camera and sent to them through www.fileflyer.com website using computer and internet facilities. The advises were given by the Scientists to the FC using mobile phone which were transferred to the farmers in written form by the FC on regular basis without involving any cost to the farmer. Both soil and water sampling were done for all the 25 farms to analyze the soil and water status and the recommendations for the problems were offered considering both the digital images and these sampling results.

Similar model of e-Velanmai was adopted at Aliyar sub basin of Coimbatore district from 1st May 2008 covering 25 farms from five villages drawn from five WUA. The same model of e-Velanmai was replicated at Varaghanadhi sub basin in Villupuram district of Tamil Nadu during 1st July 2008. These two districts were distinctly different in terms of crops, water resource and agroclimatic conditions. However this model was discontinued due to poor response of the farmers to avail the ICT based extension advice offered on cost free basis in the study locale. The experiment gave a lesson that even the quality and timely extension advice when offered to the clients at free of cost on supply driven mode will not motivate the farmers to participate in the extension process.

For the purpose of ensuring the participation of farmers to benefit from the ICT based extension process, it was thought to alter the implementation strategy from public model to a paid model of e-Velanmai. The paid model was conceived and introduced first at Kumarapalayam WUA of the Palar sub basin from 18th October 2008.

Paid Model of e-Velanmai

In order to enhance the participation of farmers in scientific farming and to have sustainability of the scheme even after the project period, paid model of e-Velanmai was conceived and the same was first introduced at Palar sub basin, followed by Chendur WUA of the Varaghanadhi sub basin from 12th January 2009 and at Mannur WUA of Aliyar sub basin from 1st February 2009.

Under this model, the interested farmers of the selected WUA from the respective sub basins were enrolled as members of the e-Velanmai scheme by collecting the membership fee based on the farm size owned by them. A membership card was given to the farmer and a record of the agricultural problems and advices given by the scientists were noted in the card. These members availed the benefits of e-Velanmai extension activities as mentioned under the public model.

In each of the member's family, at least one person was thoroughly trained in handling the ICT tools for the purpose of framing the crop status images and to access the advices from experts. It was envisioned that the trained individual is expected to capture and send the digital images of the pest damaged symptoms and receive the technical advice from the experts of TNAU. The paid model of e-Velanmai has been upscaled in the following 19 subbasins (Phase I and II) during 2011-2012 (Table 1.2) and 26 sub basins (Phase I to IV) during 2012-2013 (Table 1.3). In each sub basin, one scientist took the charge of implementing the scheme. One FC was appointed to link the farmers and scientists for the purpose of technology transfer. An expert team comprising of crop protection and crop management was identified in either a KVK/Research station located in each of the sub basins to offer scientific advices to farmers.

Table 1.2: Upscaling of e-Velanmai (2011-2012)

Sub Basins	
I Phase	*II Phase*
Varaghanadhi	Ponnaiyar
Upper Vellar	Agniyar
South Vellar	Ambuliyar
Pambar	Swethanadhi
Kottakaraiyar	Nichabanadhi
Manimuthar	Kalingalar
Arjunanadhi	Therkar
Aliyar	Upper Gundar
Palar	Koundanyanadhi
	Poiney

Table 1.3: Upscaling of e-Velanmai (2012-2013)

Sub Basins	
I – PHASE	Nallavur
Varaghanadhi	Ongur
Upper Vellar	Thurinjalar
South Vellar	Gadilam
Kottakaraiyar	Gridhamal
Manimuthar	Uthirakosamangaiyar
PAP - Palar	Deviyar
II – PHASE	Hanumanadhi
Ponnaiyar	Kambainallur
Agniyar	*IV – PHASE*
Swethanadhi	Cooum
Koundinyanadhi	Adayar
III – PHASE	CheyyarKillyar
Araniyar	Paralaiaru
Kosasthalaiyar	Amaravathi

Method of Technology Transfer

The operational procedure in the technology transfer process (Figure 1.1) was initiated by any farmer who is in need of technical advice in the sub basin. The farmer calls the FC over his mobile and the FC goes to the farmer's field and examines his crop status. In case the farmer is in need of DB advices such as suitable variety, marketing decisions, spacing *etc.*, then the advices were given on the spot by the FC by referring to the TNAU agri portal. In the case of a PB query raised by the farmer say for example the management of an insect pest, disease, nutritional

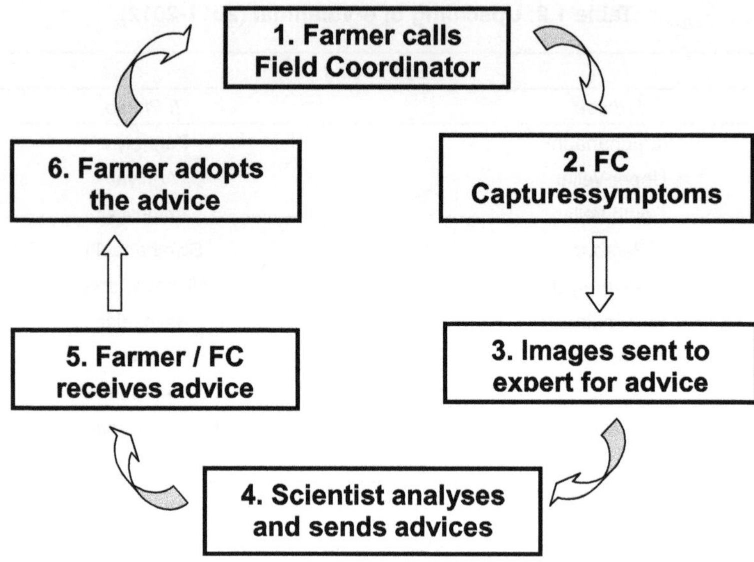

Figure 1.1: Stages in Technology Transfer Process in e-Velanmai (2007-13).

deficiency, weeds *etc.*, the advices will be given to the farmer based on a digital photo document captured by the FC and sent to the experts using internet. The images were coded using the farmer's name and identity number. The experts diagnosed the problem photos and offered solutions to the respective cases. These recommendations were downloaded and recorded in the membership card given to the farmers respectively.

Technology Transfer Process using Android App integrated with FCMS website

During 2014-15, an android app named 'e-Velanmai-FCMS' app was created and the same is integrated with the Farm Crop Management System (FCMS) website of the Department of Agriculture(DOA), Government of Tamil Nadu to ease the technology transferprocess attempted by the Extension Officials (Block Technology Managers and Assistant Technology Managers of ATMA). This app can be downloaded in a tablet media and operated by Extension Officials of the DOA and Scientists of TNAU to render technology advisory services to farmers.

Results and Discussion

The results on the experimentation (Pilot test +Upscaling) of the scheme and the lessons learnt are presented as follows:

Membership Status in e-Velanmai

Farmers were enrolled in the scheme as members by collecting a membership fee of ₹50 to 150 per farmer based on their farm size. About 10507 farmers had joined the scheme (Table 1.4) by paying the prescribed annual membership fees. The gender wise distributions among the members revealed that 89.76 per cent were male and the rest were female. The farm size wise distribution of the members indicated that a majority of small and marginal farmers (80.60 per cent) who operated < 5 ac land holding had participated largely as compared to medium and big farmers. The fact that more than three-fourth proportions of farmers in the state were small farmers could be attributed for the result obtained. The results suggest that farmers irrespective of their land holding had expressed their willingness to pay behavior towards availing quality and timely extension advices using ICT tools to adopt scientific practices in their farms.

Farmers' Contribution to Avail Extension Services

Depending on the farm size owned by the farmers, they paid the membership fee of ₹50 to ₹150 per farmer. The members were offered extension advices and training on handling ICT tools. A total amount of ₹6,54,700 was mobilized from the farmers to offer quality and timely advices from the experts of TNAU. An amount of ₹5,94,050 was contributed by male farmers and the share of farm women was ₹60,650 (Table 1.5).

This amount was deposited in rural banks by creating a separate account in the name of the social organization/WUA maintained as joint account to be operated by the leader of the social organization and the scientist-in-charge of the sub basin. The money was proposed to be used to avail extension advices/agril. information from the experts/media sustainably after the closure of the scheme in the sub basin.

Agricultural Extension Advisory Services

Experiences on answering the queries raised by the farmers revealed that there existed two types namely (i) Problem based (PB) queries and (ii) Decision based (DB) queries. PB queries were answered with the help of the crop status images (*e.g.* Pest damaged symptoms) whereas the DB queries (*e.g.* Marketing decisions) were answered on the spot using the TNAU agriportal website and other technical data base in agriculture. These queries were also analysed according to the subject matter it represented. The results reveal that about 20,259 queries were tackled by offering suitable technical advices to the farmers. Among these queries, a majority come under the plant protection subject (54.22 per cent) followed by crop management (34.36 per cent) and other subjects (11.42 per cent) such as Marketing, Post harvest technologies, Agricultural Engineering *etc.* It is a well known fact that the presence of an insect pest or disease or lack ofmicro and macro essential nutrients in a crop will express damage symptoms and hence these problems were visible and makes the farmer to get concerned about the problem. This might have influenced the farmers to raise more PB queries (73.97 per cent) focussed on crop protection aspects followed by management DB queries (26.03 per cent). It is proved that an agricultural expert can offer effective technical advices to farmers based on the digital images of the crop status, even without visiting the farmers field in person (Table 1.6).

Table 1.4: Membership Enrollment in e-Velanmai

26 Sub Basin	*No. of Farmers Enrolled*									*Total No. of Members Enrolled*		
	<5 ac			*5-10 ac*			*>10 ac*			*Male*	*Female*	*Total*
	Male	*Female*	*Total*	*Male*	*Female*	*Total*	*Male*	*Female*	*Total*			
Total	7490	979	8469	1814	93	1907	126	4	130	9431	1076	10507
Per cent	71.29	9.32	80.60	17.26	0.89	18.15	1.20	0.04	1.24	89.76	10.24	100

Table 1.5: Membership Fees Contributed by Farmers

Sub Basin	*No. of Farmers Enrolled*									*Total No. of Members Enrolled*		
	<5 ac			*5-10 ac*			*>10 ac*			*Male*	*Female*	*Total*
	Male	*Female*	*Total*	*Male*	*Female*	*Total*	*Male*	*Female*	*Total*			
Total	374900	49300	424200	189500	10300	199800	29650	1050	30700	594050	60650	654700
Per cent	57.26	7.53	64.79	28.94	1.57	30.52	4.53	0.16	4.69	90.74	9.26	100

Table 1.6: Technologies Transferred to Farmers

Year	*Subject-wise No. of Advices given*								*Total No. of Advices given*		*Grand Total*
	Pathology	*Ento-mology*	*Physiology*	*Soil Science*	*Agronomy*		*Marketing*	*Others*			
	PB	*PB*	*PB*	*DB*	*PB*	*DB*	*DB*	*DB*	*PB*	*DB*	
Advices	5643	5343	1577	496	764	4117	1301	1028	14985	5274	20259
Per cent	27.85	26.37	7.78	2.49	3.77	20.32	6.42	5.07	73.97	26.03	100

Adoption Status of the Advices Offered to Farmers

The farmers' queries were answered by the experts in the case of PB queries and the DB queries were answered on the spot by the FC or after getting the advice from the experts. These advices were adopted at farm level by an overwhelming majority of the farmers. The ratio between the number of farmers queried to adopted was found to be 84.61 per cent (Table 1.7). The success rate of the adopted farmers was found to be 91.10 per cent and the remaining seems to be landed in failure due to various reasons such as shortage of labourers for adoption of recommended practices, lack of conviction of farmers over the recommended practices and high cost of recommended inputs.

Table 1.7: Adoption Status of the Advices Offered to Farmers

26 Sub basins	*Number of Farmers Queried*	*Number of Farmers Adopted*	**Success Rate (Per cent)*
Total	20259	17142	91.10

Rate of adoption 84.61 per cent.

*(Success rate due to the adoption of advices among the adopted farmers).

Reasons for Non-adoption of the Recommended Advices

The problems confronted by the farmers were identified either by the Scientist or the FC and the appropriate advices were recommended to the farmers in time within 3 hours. However some farmers did not adopt the advices in their farm due to various reasons. Hence a survey was collected with 500 randomly selected samples from all the selected sub basins and the results are given in the Table 1.8.

Table 1.8: Reasons for Non-adoption of the Recommended Advices

Sl.No.	*Reasons for Non-adoption of Advices*	*No. of Farmers* (n=500)*	*Percentage*
1	Non-availability of prescribed pesticides in local area	118	23.60
2	Shortage of labourers for adoption	103	20.60
3	Lack of conviction of farmers for want of immediate observation of results	85	17.00
4	High cost of recommended inputs	68	13.60
5	Resistance to change attitude of farmers	35	7.00

* Multiple responses recorded.

The findings on the reasons for non-adoption of the recommended advices revealed that a majority of the farmers (23.60 per cent) felt that they were not able to procure the recommended agro inputs from the local input shops. Shortage of labour to perform various agricultural operations in the village was also identified as one of the reason for non-adoption (20.60). The other reasons for non-adoption of the advices include, high cost of the recommended inputs (13.60 per cent) such as Methomyl, Nativo *etc.*, which was sold for ₹1000 to 6000 per kg. Lack of conviction of farmers for want of observable results in the field immediately after adoption of the practices (17 per cent), resistance to change attitude of farmers (7 per cent) for adoption of new technologies.

Reasons for Failure After Adoption of the Advices

The technologies advised to the farmers were not adopted scientifically as per the correct dose and specifications recommended. Hence 9 per cent of the farmers reported failure in their farms (Table 1.9). Agricultural operations have to be done in time to achieve a successful crop. However depending upon the availability of the labour in the village as well as the sprayer availability on hire basis from the owner farmers, the spraying of pesticides was undertaken by the farmers which had not favoured the farmers for a successful crop. Hence this had resulted in failure as reported by 7.6 per cent of the farmers. Some farmers (4.4 per cent) opined that the input shop keepers interfered in altering the suggestions of the experts and attempted to recommend higher dosage of inputs/alternate inputs thus preventing the farmers to adopt as per the advices of the experts which had not yielded good results in the farm. Few farmers (3.6 per cent) received the recommendation in time but attempted to adopt latter thus missing the appropriate growth stage of the crop leading to failure. Other group of farmers had reported that they had ignored the recommended seed treatment technology in paddy due to sudden onset of monsoon in their village thus landing in the failure of the crop.

Table 1.9: Reasons for Failure after Adoption

Sl.No.	*Reasons for Failure After Adoption of Advices*	*No. of Farmers (n=500)*	*Percentage*
1	Improper adoption against recommendation	45	9.00
2	Spraying based on the labour/sprayer availability	38	7.60
3	Interference of the private dealer in adoption of the recommendation	22	4.40
4	Adopting the recommendation at inappropriate stage of the crop.	18	3.60
5	Urgency in sowing operation ignoring the seed treatment technology	18	3.60

Benefits/advantages realized by farmers due to e-Velanmai model of extension

A random sample of 500 farmers from the study locale were enquired about the benefits they realized due to the adoption of technical advices received through the e-Velanmai method of extension. The results are presented in the Table 1.10.

It was observed that an overwhelming proportion of farmers felt that they were able to receive the advices timely on the same day when they raised the query to the field coordinator of the project. They were also happy to enjoy the benefit of access to the recommendations either at their door steps/farm level itself. Nearly three-fourth of the beneficiaries opined that they were able to achieve input use efficiency *i.e.*, saved money on investment over inputs use to manage the crop production or protect the crop from various pest and diseases. Almost an equal proportion of farmers reduced their crop loss due to the adoption of the recommended management practices resulting in achievement of higher yield and income. More than half of the beneficiaries reported that they were able to adopt the precise and appropriate recommendations to tackle various problems confronted in their farm.

Table 1.10: Benefits/Advantages Realized by Farmers

Sl.No.	*Benefits Realized*	*No. of Farmers* (N=500)*	*Percentage*
1	Timely advices were received at individual's farm level/ door step	455	91
2	Saved input cost to manage/protect a crop	375	75
3	Reduced crop loss/obtained higher yield and earned more income.	365	73
4	Precise and appropriate recommendations adopted to tackle problems	310	62
5	Dependence over input dealers for advices decreased	240	48
6	Adopted improved/scientific methods to manage the pest and diseases	180	36
7	Gained awareness/knowledge about the management of various pests/symptoms/and fertilizer dosage.	165	33
8	Use of biofertilizers, biocontrol methods for different crops were learnt through FC	125	25
9	Awareness/acquired skill to operate ICT tools like camera and internet for further contact with experts	125	25
10	Functioning of social group through e-velanmai scheme	75	15

* Multiple responses recorded.

Prior to the introduction of e-Velanmai project in the village, nearly half of the farmers stated that they depended on the input shop keepers/dealers available in their locale for crop protection advices. This situation had changed for them after the introduction of e-Velanmai as they were able to access all advices related to agriculture at their farm gate itself. About one-third of the farmers reported that they were able to adopt improved management/crop protection practices. In other words, one-third of the farmers stated that they had come to know about various pests, their symptoms and management technologies in agriculture.

About one-fourth of the beneficiaries expressed that they had learnt about the use of bio-fertilizers and bio-control methods in crop protection from the field coordinators as such technologies were recommended to the farmers. Farmers also reported that they had acquired skill to operate the ICT tools such as digital camera to capture crop status images or internet to access the recommendations from the experts of the project. Few farmers had opined that they gained social recognition and support to share agricultural information among a group of farmers due to the functioning of social groups formed under the e-Velanmai project at their village level.

It could be inferred that farmers had endorsed several economic, personal and social benefits/advantages due to their participation in the e-Velanmai project to access and adopt scientific advices to tackle their farm related problems.

Experiments on Sustainability of 'e-Velanmai'

The sustainability of offering the extension advisory services through 'e-Velanmai' approach even after withdrawal of support from TN-IAMWARM project was studied. PP model to sustain e-Velanmai was experimented and the results are presented as follows.

Public-Public Partnership

It was felt that the DOA which is holding the major responsibility to transfer agricultural technologies to farmers in the state could be an option to sustain the project. Accordingly, the Joint Director of Agriculture, Coimbatore district and the Deputy Director of Agriculture (DPAP), Coimbatore were approached to involve the Block Technology Manager (BTM) working under ATMA of one block coming under Palar sub basin namely the Sultanpet block to enact the role of Field Coordinator (FC) and offer ICT enabled extension advices to the farmers. The officials accepted and gave instructions to the BTM for offering the extension advices to farmers using the already available data base regarding the crop status and recommendations offered to the Palar sub basin farmers. An awareness meeting to the farmers about the involvement of the BTM as FC was held on 26th July 2011 at Senjerimalai village. Farmers accepted the idea to obtain the e-Velanmai mode of technology transfer through BTM instead of FC engaged in the project.

It was found that the BTM of Sultanpet block is capable to transfer appropriate technologies to farmers using the existing data base on crop status images and advices developed under the project. It was also understood that BTM/SMS could be involved effectively to offer ICT enabled agricultural extension services to farmers and the project could be operated sustainably in Tamil Nadu. Considering the socio-economic scenario of farmers in Tamil Nadu, *it is stated that Public – Public partnership could be a viable and feasible option to sustain e-Velanmai project in Tamil Nadu,* for which suitable orders were also issued latter by the Government of Tamil Nadu.

Lessons Learnt

The lessons learnt from the action research project is listed as follows:

1. Appropriate farm specific and timely agricultural extension advices can be offered by the agricultural experts to solve the farmer's problems based on the crop status images sent to by field extension staff or farmers from the field using ICT tools.
2. The limitations of physical distance that exist between the expert and farmer besides inadequate availability of transportation facilities for performing extension advisory services could be overcome by e-Velanmai method of technology transfer.

3. Farmers will come forward to pay a nominal fee based on their farm size for accessing quality and timely extension advices from agricultural experts.
4. Most of the farmers were willing to pay for accessing agricultural advices through ICT mode like e-Velanmai. Hence, farmers were ready to pay for accessing quality advices which are rendered timely by the experts.
5. Farmers across gender and farm size had shown interest and participated in the ICT enabled extension process for adoption of scientific farming.
6. The upscaling of e-Velanmai project is feasible, workable and replicable in the irrigation project command areas of Tamil Nadu.
7. It is possible to train at least one member of the farmers' family to handle ICT tools to access scientific recommendations from the experts.
8. It is possible to sustain the ICT enabled personalized extension approach by involving the BTM/ATM of ATMA as FC for technology transfer to farmers.

Adoption and Implementation of e-Velanmai by the State Department of Agriculture (DOA) (2015-16)

The World Bank experts took the evaluation of e-Velanmai project periodically and recommended the extension approach for adoption by the DOA in 100 Blocks of 13 districts in Tamil Nadu state. Suitable Government orders (Vide G.O.4 (D). No. 11 Dt.15.05.2014 Agriculture (WD2) Department of the Agril. Production Commissioner and Secretary). An amount of ₹50 Lakhs was sanctioned to execute the project jointly by the Tamil Nadu Agricultural University and DOA during 2015-16.

e-Velanmai Extension Approach Implemented during 2015-16

An android app. for performing e-Velanmaiadvisory services was created under the project and integrated with the Farm Crop Management System (FCMS) software of the DOA. The e-Velanmai-FCMS app can be loaded in any mobile/ tablets with android platform. The extension officers, specifically the BTM and ATM working under ATMA project were engaged to perform the extension advisory services to the farmers in the 100 Blocks. The extension officials were provided with tablet loaded with e-Velanmai-FCMS android app. About 200 Assistant Directors of Agriculture and Agricultural Officers working in the 100 blocks were trained on using the app to execute technology transfer services to the farming community during June – July of 2015. These master trainers will in turn train the BTM/ATMs on e-Velanmai method of extension who are working under their control in the 100 blocks of Tamil Nadu.

Stages in Technology Transfer using e-Velanmai Android App (2015-16)

The BTM/ATM of each block will visit the farmer's field in their block and capture the problematic crop status images along with the farmer's mobile number. These images will be sent to the Experts at Level I (Agricultural Officer of the

Block) by opening the log-in page of the e-Velanmai android app. using the pass word and user name assigned for a particular block. The experts at Level I will open the advisory page of the e-Velanmai-FCMS app. in the tablet to diagnose the symptoms of the pest and send the SMS - advices in Tamil version to the farmer whose crop is affected by the pest. In case, if he is unable to identify the pest/give a recommendation he is at freedom to export the crop status image to the expert at Level II (TNAU Scientist) seeking an appropriate advice for the farmer. The team of experts at Level II would use their codes to open the e-Velanmai-FCMS app. to offer recommendation to the farmers for tackling the pest in his crop. The advices will be prepared in Tamil version and sent to the farmer and extension officer who has sent the image to the scientist. The SMS services in Tamil version were offered at free of cost by accessing the Kisan SMS portal of the GOI.Later during subsequent visits the BTM/ATM visited the farmer's field to check the adoption status.

During Aug-Dec., 2015, using the android app. about 2569 advisories were offered to the farmers covering various crops and pest types by the agricultural experts. Among the crops, Cereals and Vegetables growers received majority of the advices to manage disease and insect pest problems (Table 1.11). The major reason for the distribution of advisories among the crop and Pest type might be due to the area occupied by the crops and the importance given by the farmers to manage the visible symptoms of the Pest in the field.

Table 1.11: Crop-wise Number of Advisories Offered to Farmers during 2015

Pest/ Crop Type	*Cereals*	*Pulses*	*Oilseeds*	*Vegetables*	*Cash Crops*	*Total No. of Advices*
Diseases	377	157	154	177	132	997
Insect pest	245	168	173	142	110	838
Nutritional Disorder	56	23	56	59	35	229
Weeds	95	28	34	46	67	270
Fertilizers	23	35	13	34	10	115
Marketing and others	61	48	59	63	49	120
Total No. of Advices	857	459	489	521	403	2569

A random sample of 60 Extension Officials (BTM) were contacted to understand the perceived advantages about the e-Velanmai-FCMS app. The results are presented in Table 1.12. which indicate that they were successful in convincing the farmers to adopt the recommended advices as they were able to highlight various advantages that could be obtained due to the adoption of the advices (83.33 per cent). More than three-fourth of them felt that they were able to advice the farmers instantly using the app as it had the facility to pop up the appropriate recommendations for the pest identified by them (76.66 per cent). About 63.33 per cent of the extension officials perceived that the app is user-friendly to deliver appropriate recommendations to solve the technological needs of farmers. About half of them (51.66 per cent) felt that they had the opportunity to update their knowledge on pest management. More than one-third felt that they were able to win the confidence of farmers due to appropriate advices delivered at the right time to them. It can be inferred that the extension officials had perceived positively and felt that it is a usefull tool for technology transfer to farmers.

Table 1.12: Perception of Extension Officials (BTM/ATM) about the e-Velanmai–FCMS App

Sl.No.	*Perceived Advantages*	*Frequency (per cent) (n=60)*
1.	Farmers get convinced to adopt the SMS advices due to personal contact	50 (83.33)
2.	Can provide advices instantly to farmers	46 (76.66)
3.	User-friendly app to offer advisories in agriculture to farmers	38 (63.33)
4.	Gained knowledge about advisories on pest management	31 (51.66)
5.	Able to win over the confidence of farmers	29 (48.33)

Conclusion

The e-Velanmai approach of extension has been successfully pilot tested, upscaled and further implemented for adoption by the extension officials of the DOA in Tamil Nadu. This extension approach is found to be reliable and valid one to disseminate cost-effective and timely advisory services to the needed farmers over mobile phones in the form of SMS advices and followed by an extension personnel (BTM/ATM) to clarify and guide the technology adoption process. This does not involve huge investment and technical manpower to execute at field level. The basic infrastructural requirements for implementing this project are mobile app, tablet with internet to be possessed by the experts. Hence, this extension model is applicable for adoption/implementation in all the states of India as well as all developing countries of our world.

References

Karthikeyan, C. 2012. "e-Velanmai" – An ICT Enabled Agricultural Extension Model. **International J. of Extension Education**, Vol.8: pp. 24-30.

Mishra, D. C., 1996. "Agricultural extension effectiveness in India. A cross sectional analysis", **Indian Journal of Extension Education**, Vol. 32, No.1-4.

that they had the opportunity to update their knowledge on pest management. More than one third felt that they were able to win the confidence of farmers due to appropriate advices delivered at the right time to them. It can be inferred that the extension officials had perceived positively and felt that it is a useful tool for technology transfer to farmers.

Table 1.12: Perception of Extension Officials (BTM/ATM) about the e-Velanmai-PCMS App

Sl.No.	Perceived Advantages	Frequency (per cent)
[illegible]	[illegible]	[illegible]
[illegible]	[illegible]	[illegible]
[illegible]	[illegible]	[illegible]
[illegible]	[illegible]	[illegible]

Chapter 2

Integrated Water Management for Natural Resource Conservation

B.J. Pandian

The Director, Water Technology Centre,
Tamil Nadu Agricultural University,
Coimbatore – 641 003, Tamil Nadu

Introduction

Water is considered as the most critical resource for sustainable development in most of the countries. It is essential not only for agriculture, industry and economic growth, but also it is the most important component of the environment, with significant impact on health and nature conservation. Currently, the rapid growth of population, together with the extension of irrigation, industrial development and the climate change, are stressing the quantity and quality aspects of the natural system.

Global irrigated area has increased more than six fold over the last century, from approximately 40 million hectares in 1900 to more than 260 million hectares (FAO, 1999). Today 40 per cent of the world's food comes from the 18 per cent of the cropland that is irrigated. Irrigated areas increase almost 1 per cent per year and the irrigation water demand will increase by 13.6 per cent by 2025 (Rosegrant and Cai, 2002). On the other hand 8-15 per cent of fresh water supplies will be diverted from agriculture to meet the increased demand of domestic use and industry. Furthermore the efficiency of irrigation is very low, since only 55 per cent of the water is used by the crop (Figure 2.1). To overcome water shortage for agriculture is essential to increase the water use efficiency and to use marginal waters (reclaimed, saline, drainage) for irrigation.

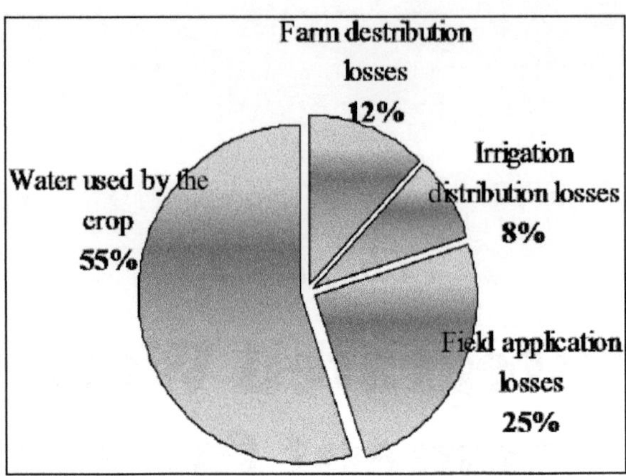

Figure 2.1: Water Losses in Agriculture.

Water and Agricultural Production

Agriculture currently uses about 70 per cent of the total water withdraw, mainly for irrigation. Although irrigation has been practiced for millennia, most irrigated lands were introduced in the 20th century. The intensive irrigation could provide for the growth of irrigated areas and guarantee increased food production. In the 1980s, the global rate of increase in irrigated areas slowed considerably, mainly due to very high cost of irrigation system construction, soil salinization, the depletion of irrigation water-supplying sources, and the problems of environmental protection. However, as the population is growing at a rapid rate, irrigation is being given an important role in increasing land use and cattle-breeding efficiency. Thus, irrigated farming is expected to expand rapidly in the future with subsequent increase of water use for irrigation.

Irrigation is not sustainable if water supplies are not reliable. Especially in areas of water scarcity the major need for development of irrigation is to minimize water use. Effort is needed to find economic crops using minimal water, to use application methods that minimize loss of water by evaporation from the soil or percolation of water beyond the depth of root zone and to minimize losses of water from storage or delivery systems. Nowadays, during a period of dramatic changes and water resources uncertainty there is a need to provide some support and encouragement to farmers to move from their traditional high-water demand cropping and irrigation practices to modern, reduced demand systems and technologies.

Under scarcity conditions considerable effort has been devoted over time to introduce policies aiming to increase water efficiency based on the assertion that more can be achieved with less water through better management. Better management usually refers to improvement of allocative and/or irrigation water efficiency. The former is closely related to adequate pricing, while the latter depends on the type of irrigation technology, environmental conditions and on scheduling of water application.

Over-irrigation can cause among others temporal water shortage to other farmers, water-logging conditions for the crop, favorable environment for disease development, loss of nutrients due to leaching or deep percolation, contamination of the aquifers from agrochemicals, reduction of crop yield and deterioration of the quality and increase of production cost.

Integrated Water Managment in Agriculture

Sustainable water management in agriculture aims to match water availability and water needs in quantity and quality, in space and time, at reasonable cost and with acceptable environmental impact. Its adoption involves technological problems, social behavior of rural communities, economic constrains, legal and institutional framework and agricultural practices.

Under water demand management most attention has been given to irrigation scheduling (when to irrigate and how much water to apply) giving minor role to irrigation methods (how to apply the water in the field). Many parameters like crop growth stage and its sensitivity to water stress, climatic conditions and water availability in the soil determine when to irrigate or the so-called irrigation frequency. However, this frequency depends upon the irrigation method and therefore, both irrigation scheduling and the irrigation method are inter-related.

Localized Irrigation Methods

Localized irrigation is widely recognized as one of the most efficient methods of watering crops. Localized irrigation systems (trickle or drip irrigation, micro-sprayers) apply the water to individual plants by means of plastic pipes, usually laid on the ground surface. With drip irrigation water is slowly applied through small emitter openings from plastic pipes with discharge rate ≤ 12 l/h. With micro-sprayer (micro-sprinkler) irrigation water is sprayed over the part of the soil surface occupied by the plant with a discharge rate of 12 to 200 l/h. The aims of localized irrigation are mainly the application of water directly into the root system under conditions of high availability, the avoidance of water losses during or after water application and the reduction of the water application cost (less labour).

The main characteristics of localized irrigation methods:

- ☆ **Low rate of water application**(discharge rate < 200 l/h for mini sprinklers, <12 l/h for drippers and an application rate 1-5 mm/h)
- ☆ **Partial soil wetting:**Wetted soil is a portion of the soil volume available to the roots, 30-40 per cent for tree crops, and 50-80 per cent for vegetables.
- ☆ **Low doses, high frequency, long duration of irrigation:** Doses 1/3 - 1/10 of those used for surface methods. High frequency (usually one irrigation per 1-7 days)
- ☆ **High soil water availability:** Slow and frequent irrigations ensure that water content in the soil remains high and fairly constant and the soil water tension remains low (1/3 of atm) resulting in high water availability to the plant.

Drip irrigation reduces water use by 30 to 70 per cent and raises crop yields by 20 to 90 per cent.

The Advantages of Localized Irrigation Method

- ☆ Efficient water use (efficiency up to 95 per cent)
- ☆ Reduced labour cost (can be highly automated)
- ☆ Easy and efficient application of fertilizer and other chemicals
- ☆ Reduced salinity hazards
- ☆ Better phytosanitary conditions
- ☆ Simultaneous performance of other cultural practices
- ☆ Can be used on uneven or sloping areas and
- ☆ Valorization of small water discharges.

Drip irrigation's combination of water savings and higher yields typically increases at least by 50 per cent the water use efficiency, yield per unit water, and makes it a leading technology in the global challenge of boosting crop production in the face of serious water constrains.

The main components of a drip irrigation system (Figure 2.2)

1. The pressure source (collective pressurized network, pumping station or elevated water storage tank)

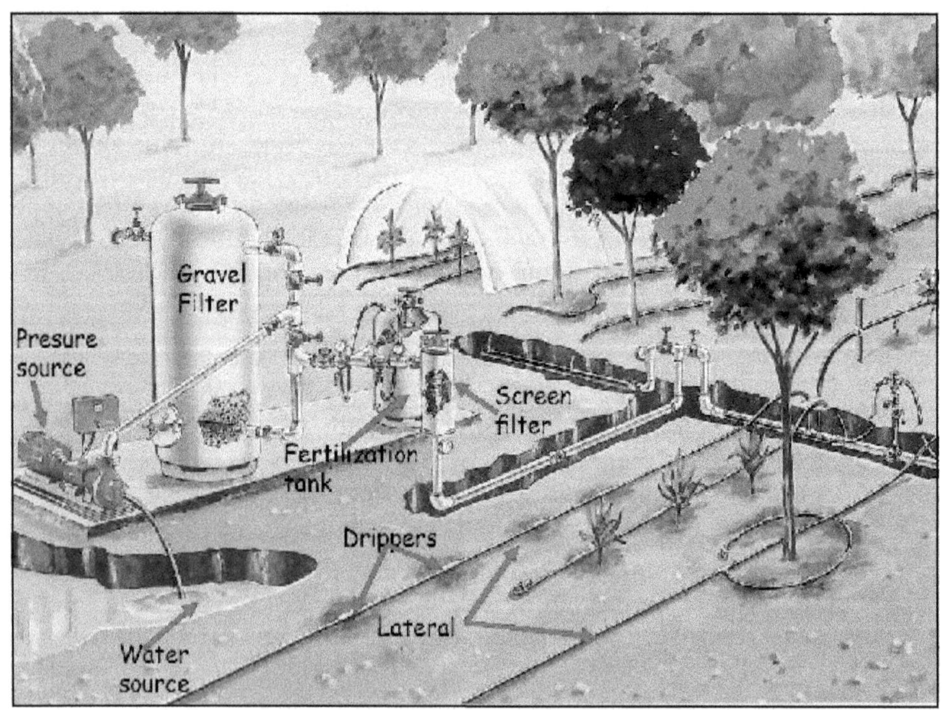

Figure 2.2: Layout of Drip Irrigation System.

2. The control head (different kinds of filters, fertilizer tank, pressure regulators, valves, water meters, automation devices)
3. The pipelines (main pipe, manifolds and laterals) and
4. The emitters (drippers –'on line', in line', self-cleaning or auto-regulated, mini-sprayers – stable or rotating).

Improvements in localized irrigation systems aiming to reduce the volumes of water applied and increase the water productivity include the use of a single drip line for a double row crop, the use of micro-sprayers in high infiltration soils, the adjustment of duration of water application and timing to soil and crop characteristics, the control of pressure and discharge variations, the use of appropriate filters to the water quality and the emitter characteristics used, the adoption of careful maintenance, automation, fertigation (efficient fertilizer application) and chemigation (easy control of weeds and soil born diseases).

Irrigation Scheduling

Irrigation scheduling is the decision making process for determining when to irrigate the crops and how much water to apply. It requires good knowledge of the crops water requirements and of the soil water characteristics that determine when to irrigate, while the adequacy of the irrigation method determines the accuracy of how much water to apply. In most cases, the skill of the farmer determines the effectiveness of the irrigation scheduling at field level. With appropriate irrigation scheduling deep percolation and transport of fertilizers and agro-chemicals out of the root-zone is controlled, water-logging is avoided, less water is used (water and energy saving), optimum soil water conditions are created for plant growth, higher yields and better quality are obtained and rising of saline water table is avoided. In water scarce regions, irrigation scheduling is more important than under conditions of abundant water, since any excess in water use is a potential cause for deficit for other users or uses.

Irrigation scheduling techniques and tools are quite varied and have different characteristics relative to their applicability and effectiveness. Timing and depth criteria for irrigation scheduling can be established by using several approaches based on soil water measurements, soil water balance estimates and plant stress indicators, in combination with simple rules or very sophisticated models. Many of them are still applicable in research or need further developments before they can be used in practice. Most of them require technical support by extension officers, extension programmes and technological expertise of the farmers. However, in most countries these programmes do not exist because they are expensive, trained extension officers are lacking, farmers awareness of water saving in irrigation is not enough and the institutional mechanisms developed for irrigation management give low priority to farm systems. Therefore, in general, large limitations occur for their use in the farmers practice.

Soil Water Estimates and Measurements

Soil water affects plant growth directly through its controlling effect on plant water status. There are two ways to assess the availability of soil water for plant growth by measuring the soil water content, and by measuring how strongly that water is retained in the soil (soil water potential). The accuracy of the information relates to the sampling methods adopted and to the selection of locations where point observations are performed due to the soil water variability both in space and depth. Soil water estimates and measurements used for irrigation scheduling include:

a. Soil Appearance and Feel

Assessment of the soil water status by feeling how dry is the soil using the hands or a shovel. It is applicable mainly to field crops than to fruit trees and its effectiveness depends on farmer's experience in sensing the changes in the soil moisture by hand touching the soil.

b. Soil Water Content Measurement

It can be done through soil sampling for laboratory analysis, or using neutron probe or time-domain reflectometer (TDR). Neutron probes use the property of scattering and slowing down neutrons by the hydrogen nuclei of the water molecules. The TDR measures the propagation of an electromagnetic wave through the soil. The characteristics of this propagation depend on the soil water content through the dielectric properties of the soil. The main advantages are that non-destructive and direct measurements can be performed without disturbing the soil allowing one to follow water content changes with time. The main limitation for neutron probe relates to safety rules, which have to be followed to operate, transport and store the probe, while for TDR seem to be due to gaps and cracks which may arise during installation of the rods or as a result of shrinking of the soil during drying. Assessment of the soil water content can be done using porous blocks or electrode probes by sensing the changes in the electrical resistance in the soil due to variations in soil moisture. The above mentioned techniques are precise, their applicability is large but are expensive and require calibration and expertise or external support to farmers. Their effectiveness depends upon the selected irrigation thresholds.

c. Soil Water Potential Measurement

Tensiometers, soil psychrometers and pressure transducers are highly precise instruments for measuring soil water potential. Tensiometers, which assure low cost, simple operation and provide information for precisely determining the irrigation timing and depths when irrigation thresholds are well established, widely used for the irrigation of horticultural crops. There should be at least one, and preferably two, tensiometer locations (two or more tensiometers at one location being a station) for each area of the field that differs in the soil type and depth (Figure 2.3). The depth of tensiometer installation in the soil varies with the crop type (Table 2.1), while advise to framers is desirable. The only limitation is that they not operate beyond their sensing ability (0-80 kPa). Soil spectrometers and pressure transducers are complex, expensive and require highly qualified farmers, so their farm application is very limited.

Table 2.1: Depth of Tensiometer Installation for Various Crops

Plant Type	*Effective Root Depth (cm)*
Grasses	15
Vegetables	15-30
Shrubs	30-45
Trees	30-60

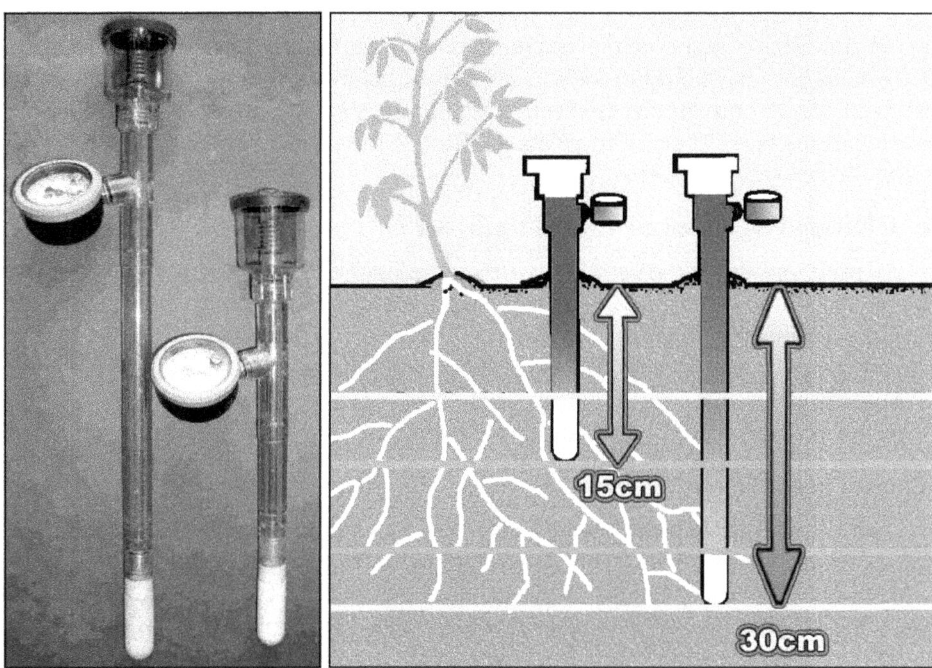

Figure 2.3: Schematic Representation of Tensiometers and their Proper Installation in the Field.

d. Remotely Sensed Soil Moisture

Data on soil moisture, mainly for sallow soil layer, are performed with an airborne thermal infrared scanner, and usually refers to large areas. Its applicability is very limited and further development is required.

Crop Stress Parameters

Instead of measuring or estimating the soil water parameters, it is possible to get messages from the plant itself indicating the time of irrigation but not defining the irrigation depths. This message can either come from individual plant tissues, then it will be necessary to have a correct sampling, or from the canopy as a whole. Therefore, crop stress parameters are useful when irrigation depths are predefined and kept constant during the irrigation season.

a. Plant Appearance

Leaf rolling, changes in leaf orientation or colour are signs in plants developing water stress, so can be used for determining the time of irrigation. However, this technique is applicable to plants that give detectable signs of water stress prior to wilting and its effectiveness depends upon farmer's experience.

b. Leaf Water Content and Leaf Water Potential

The main interest of these measurements lies in the possibility of linking values of pre-dawn leaf water potential and leaf water content to relative evapotranspiration. However, both have limitations such as representative sampling; require relatively sophisticated equipment (pressure chamber, psychrometer), timeliness for observations and expertise in interpretation of the results. Thus, they are mainly used for research purposes.

c. Changes in Stem or Fruit Diameter

Trend in diameter growth and diurnal changes of stems or fruits are measured using micrometric sensors. They are not difficult to install, mainly on tree branches, and are connected to a logging system. The main problem encountered is that, sometimes, the same response is obtained with an excess and a lack of water. Furthermore, this technique is expensive, requires expertise and well-selected thresholds, so that its use by farmers is limited.

d. Sap Flow Measurement

Sap flow measurements using appropriate electronic sensors are sometimes included among the plant water stress criteria for irrigation scheduling. In fact, one can obtain relative ET values by measuring sap flow along the trunk and comparing trees under water shortage to well irrigated trees. The techniques available are the sap flux density technique and the mass flux technique. The former is limited by the need to determine the cross-sectional area of the water conducting tissue. The latter is restricted to estimation of small trees' transpiration. Both techniques need tree sampling, require expertise, well defined thresholds and mainly used for research.

e. Canopy Temperature

This technique is applicable to field crops for large areas. It is based on the fact that the surface temperature of a well-watered crop is some degrees lower than the air temperature, while the surface temperature of a stressed crop is close to the air temperature. Surface temperature measurements are performed using infrared thermometers, while the determination of crop water stress index (CWSI) is necessary. CWSI can be used only if weather conditions are not rapidly changing (wind and radiation) and only for fully developed crops.

f. Remote Sensing of Crop Stress

Airborne and satellite scanners sensing several wavelengths including the thermal infrared are used. Information may be made available through a spatially distributed format with geographical information system (GIS). The accuracy of the technique is higher when it is applied to large cropped areas with big fields and small crop variety. It is useful to support regional irrigation programmes, while good expertise in interpreting and transmitting information to farmers is required.

Climatic Parameters

Climatic parameters are widely used for local or regional irrigation schemes. Weather data and empirical equations that, once they are locally calibrated, provide accurate estimates of reference evapotranspiration (ETo) for a given area are used. Then, crop evapotranspiration (ETc) is estimated using appropriate coefficients. Information may be processed in real time or, more often, using historical data.

Soil-water Balance

The aim of soil water balance approach is to predict the water content in the rooted soil by means of a water conservation equation: $\Delta(\text{AWC} \times \text{Root depth})$ = Balance of entering + outgoing water fluxes, where AWC is the available water content. Soil water holding characteristics, crop and climate data are used by sophisticated models to produce typical irrigation calendars. This approach can be applied from individual farms to large regional irrigation schemes. However, it needs expertise, support by strong extension services or links with information systems. Its effectiveness is very high, but depends on farm technological development and/or support services. Examples of commercial software for irrigation scheduling based on soil-water balance approach are:

IMS: Real time irrigation scheduling

MARKVAND: An Irrigation scheduling system for use under limited irrigation capacity. **SALTMED:** A computer model for generic applications that illustrates the effect of all the parameters affecting crop growth and yield.

SIMIS: A decision support system that processes information about the system soil-water-plant to provide crop water requirements and estimate irrigation needs at farm and canal level.

Effective Irrigation Scheduling

It is recognized that appropriate irrigation scheduling should lead to improvements in irrigation management performance, especially at farm level. The farmer should be able to control the timing and the depth or volume of irrigation. However, the practical application of the techniques and methods has been far below expectations. The dependence on a collective system implies social, cultural and policy constraints. The main constrains are the lack of flexibility, either due to rigid schedules or the system limitations, the non-economic pricing of water (price covers less than 30 per cent of the total cost), the high cost of irrigation scheduling (either for technology and/or labour), the lack of education and training of the framers, the institutional problems, the behavioural adaptation, the lack of interactive communication between research, extension and farmers and finally the lack of demonstration and technology transfer.

Deficit Irrigation Practices

In the past, crop irrigation requirements did not consider limitations of the available water supplies. The irrigation scheduling was then based on covering the full crop water requirements. However, in arid and semi-arid regions increasing municipal and industrial demands for water reduce steadily water allocation to agriculture. Thus, water availability is usually limited, and certainly not enough to achieve maximum yields. Then, irrigation strategies not based on full crop water requirements should be adopted for more effective and rational use of water. Such management practices include deficit irrigation, partial root drying and subsurface irrigation.

Regulated Deficit Irrigation

Regulated deficit irrigation (RDI) is an optimizing strategy under which crops are allowed to sustain some degree of water deficit and yield reduction. During regulated deficit irrigation the crop is exposed to certain level of water stress either during a particular period or throughout the growing season. The main objective of RDI is to increase water use efficiency (WUE) of the crop by eliminating irrigations that have little impact on yield, and to improve control of vegetative growth (improve fruit size and quality.) The resulting yield reduction may be small compared with the benefits gained through diverting the saved water to irrigate other crops for which water would normally be insufficient under conventional irrigation practices.

RDI is a sustainable issue to cope with water scarcity since the allowed water deficits favour water saving, control of percolation and runoff return flows and the reduction of losses of fertilizers and agrochemicals; it provides for leaching requirements to cope with salinity and the optimization approach leads to economical viability. The adoption of deficit irrigation implies appropriate knowledge of crop ET, of crop response to water deficits including the identification of critical crop growth stages, and of the economic impact of yield reduction strategies. Therefore, appropriate deficit irrigation requires some degree of technological development to support the application of irrigation scheduling techniques.

Before implementing RDI it is necessary to know the crop yield response to water (growth stage or whole period). Crop yield response for deficit irrigation is described by the equation $Y/Ym = 1\text{-}Ky\ [1\text{-}ETa/ETm]$ (Stewart *et al.*, 1977), where Y and Ym are the expected and maximum crop yield, ETa and ETm the actual and maximum ET, and Ky the crop response factor. Ky gives an indication of whether the crop is tolerant to water stress and depends on crop species, cultivar, irrigation method and growth stage. High yielding varieties are more sensitive to water stress. Crops or varieties with a short growing season are more suitable for RDI. Furthermore, in order to ensure successful RDI, it is necessary to consider the water retention capacity of the soil. In sandy soils plants undergo water stress quickly under RDI, while in deep, fine-textured soils plants have ample time to adjust to low soil water potential, and may be unaffected by low soil water content. Under RDI agronomic practices may require modifications, e.g. decrease plant population, apply less fertilizer, adopt flexible planting dates and select shorter season varieties.

RDI is applied and irrigation water contains moderate to high salt content (Na or Cl) careful monitoring of soil salinity during the RDI period is necessary and strategic leaching irrigations (every 5-7 weeks) should be applied. RDI has been applied successfully for row crops like maize, soybean, sugar beet, sunflower, potato, wheat and tree crops like citrus, olives, peaches, grapevines *etc.*

Sub-surface Drip Irrigation

Subsurface drip irrigation (SDI) is a low-pressure, low volume irrigation system that uses buried tubes to apply water. The applied water moves out of the tubes by soil matrix suction. Wetting occurs around the tube and water moves out in the soil all directions.

The Potential Advantages of SDI

1. Water Conservation
2. Enhanced Fertilizer Efficiency
3. Uniform and highly efficient water application
4. Elimination of surface infiltration problems and evaporation losses
5. Flexibility in providing frequent and light irrigations
6. Reduced problems of disease and weeds
7. Lower pressure required for operation

The main disadvantages are the high cost of initial installation and the increased possibility for clogging.

Subsurface irrigation is suitable for almost all crops, especially for high value fruit and vegetables, turfs and landscapes. A large variety of tubes are available in the market from PE tubes with built-in emitters or porous tubes that ooze water out the entire length of the tube. The tube is installed below the soil surface either by digging the ditches or by special device pulled by a tractor. The depth of installation depends upon soil characteristics and crop species ranging from 15-20 cm for vegetables and 30-50 cm for tree crops.

To avoid clogging problems it is essential for SDI to use a 200 mesh filter for most tube material. Porous tubes need more frequent flushing. However, the performance and life of any system depends on how well it is designed and operated. Back flush system must be checked at regular intervals. Water quality affects the system (high pH, salinity and iron may cause precipitates). Further problems arise if water contains organic matter, bacteria or algae. For protection of the system occasional injection of acid or acid-forming chemicals or chlorine at the end of the irrigation season help to stop precipitates. After the use of chemicals, the system has to run for a while to remove residual chemicals.

Agricultural Practices

Agricultural practices, such as soil management, watershed management, fertilizer application, disease and pest control are related with the integrated water management in agriculture and protection of the environment. Agriculture practice today is characterized by the abuse of fertilizers. Farmers very rarely carry out soil and leaf analyses in order to clarify the proper quantity and type of fertilizer needed for each crop and they apply them empirically. This practice increases considerably the cost of agricultural production and is potentially critical for the deterioration of the groundwater quality and the environment. Agrochemicals (herbicides and pesticides) are also excessively used, endangering the quality of the surface water and negatively affect the environment. Plant-protection products (pesticides) are often used preventively, even when there is not real threat in the area.

There is a large variety of traditional and modern soil and crop management practices for water conservation (runoff control, improvement of soil infiltration rate, increase soil water capacity, control of soil water evaporation) and erosion control in agriculture, some of which apply also for weed control (Pereira *et al.*, 2002). Effort should be made in their rational use of chemicals for pest and weed control in order not to further pollute the environment.

Soil Management

Soil surface tillage, which concerns shallow tillage practices to produce an increased roughness on the soil surface permitting short time storage in small depressions of the rainfall in excess to the infiltration.

Contour tillage, where soil cultivation is made along the land contour and the soil is left with small furrows and ridges that prevent runoff. This technique is also effective to control erosion and may be applied to row crops and small grains provided that field slopes are low.

Bed surface profile, which concerns cultivation of wide beds and is typically used for horticultural row crops.

Conservation tillage, including no-tillage and reduced tillage, where residuals of the previous crop are kept on the soil at planting. Mulches protect the soil from direct impact of raindrops, thus controlling crusting and sealing processes. Conservation tillage helps to maintain high levels of organic matter in the soil thus it is highly effective in improving soil infiltration and controlling erosion.

Crop Management

Several techniques have been designed to minimize the risks of crop failure and to increase the chances for beneficial crop yield using the available rainfall, such as: Selection of crop patterns taking into consideration the seasonal rainfall availability and the water productivity of the crops and crop varieties.

Adoption of water stress resistant crop varieties instead of high productive but more sensitive ones, Use of short cycle crops or varieties reduces the overall crop water requirements.

Rational Use of Fertilizers

Fertilizers (type and amount should be applied according to crop needs (based on leaf and soil analysis). Use of slow-release fertilizers if possible, is desirable.

Reducing the fertilizer rates when a low yield potential is predicted; this not only reduces costs but also minimizes the effects of salt concentration in the soil.

Application of agrochemicals in proper timing to avoid damaging beneficial organisms and integrated or biological methods to control pest and diseases, if possible.

Watershed Management

The judicious use of all the resources *i.e.*, land, vegetation and water of the watershed to achieve maximum production with minimum hazard to the natural resources and for the well being of the people.

Includes the treatment of land by using most suitable biological and engineering measures in such a manner that, the management work must be economical and socially acceptable.

Objectives of Watershed Management

1. To control damaging runoff and degradation and thereby conservation of soil and water.
2. To manage and utilize the runoff water for useful purpose.
3. To protect, conserve and improve the land of watershed for more efficient and sustained production.
4. To protect and enhance the water resource originating in the watershed.
5. To check soil erosion and to reduce the effect of sediment yield on the watershed.
6. To rehabilitate the deteriorating lands.
7. To moderate the floods peaks at downstream areas.
8. To increase infiltration of rainwater.
9. To improve and increase the production of timbers, fodder and wild life resource.
10. To enhance the ground water recharge, wherever applicable.

Components of Watershed Management Programme

I. Soil and Water Conservation

a. Tillage Practice

The loosening of soil encourages infiltration time, thus reducing runoff. While surface tillage improves soil water entry, contour cultivation arrests the runoff water and soil sediments by the ridges and furrows opened across the slope. It is estimated that shallow furrows of 8-10 cm depth opened 20-25 cm apart, can hold about 15 mm of water in the furrow depressions.

b. Summer Ploughing

In dryland soils, summer ploughing is the foremost technology to be adopted for soil and moisture conservation. When summer ploughing is employed, rain water

Soil and water conservation	Water harvesting	Crop management
• Tillage practices • Summer ploughing • Fallowing • Planting geometry • Intercropping • mulching	• Runoff inducement • Vegetation management • Surface treatment • Chemical treatment • Runoff collection • MCWH • RFWH • In- situ water harvesting • Storage and conservation • Sediment control • seepage control • Evaporation control • Recycling • Critical irrigation • Supplemental irrigation	• Crop mgt and water use • Pre-monsoon sowing • Plant population • Depth of sowing • Seed treatment

Figure 2.4: Components of Watershed Management Programme.

is stored and used by the crops during the growth period. By summer ploughing, the soil become receptive to summer showers.

c. *Fallowing for Moisture Conservation*

Traditional dryland farming systems of deep vertisols involve leaving the land fallow during rainy season and raise crops only during the post rainy season profile store moisture.

d. *Planting Geometry*

Rectangular planting was found to be more advantageous than square planting for crops raised under conserved moisture situations. Water extraction in the initial stages is usually higher with closer spacing leaving smaller amounts for future use. In wider rows, usually the inter – row contains more moisture.

e. *Intercropping*

Most intercropping systems adopt two or more crops with disparate root systems. It is a general observation that when a deep rooted crop is intercropped with a shallow rooted one, the extraction of moisture from the entire root profile is maximum with better water use efficiency.

f. *Mulching*

Mulching are of various types *viz.*, surface, stubble, soil or dust, vertical polyethylene mulches *etc.* Surface mulches will be more useful when sub surface soil has adequate water.

II. Water Harvesting

1. *Runoff Inducement*

The success or failure of rain water depends on the quality of water that can be harvested from an area under given climatic conditions. The threshold retention of a catchment is the quantity of precipitation required to initiate runoff and it depends on various components such as surface storage, rainfall intensity and infiltration capacity.

Vegetation Management

The main effect of vegetation management is more pronounced in areas where the infiltration capacity is less and surface storage more. Runoff can be increased by vegetation management in areas with an annual precipitation in excess of 280 mm.

a. Surface Treatment

Surface treatment such as rock clearing, smoothing and compacting are usually done in combination. The main effect of surface treatment is it reduces surface storage.

b. Chemical Treatments

Many chemicals have been tested for water harvesting and among these sodium these sodium salts, paraffin wax and asphalt seem to offer good prospects for future application. The principle is making the soil hydrophobic by the water repellent property. Sodium salts cause the clay particles in the soil to disperse and partly seal the pores while paraffin wax and asphalt clog the pores themselves.

2. *Runoff Collection*

a. Micro-Catchment Water Harvesting

The main aim of MCWH is to store sufficient runoff water in the root zone below the basin during rains so as to cover the water requirement of the crop during the growing season.

b. Run-off Farming Water Harvesting (RFWH)

Run-off farming water harvesting is the method of collecting surface runoff from a catchment area using channels or diversion systems and storage it in a surface reservoir or in the root zone of a farmed area for direct consumption use.

c. In-situ Water Harvesting

The effective method of in-situ water harvesting are summer ploughing, broadbed and furrow, ridges and furrows, random tie ridging, compartmental bunding *etc.* the various in-situ water harvesting methods for block and red soils cause an increase of upto 15 per cent in crop yields.

3. *Storage and Conservation*

a. *Sedimentation Control*

In deep slope runoff water carries considerable amount of surface soil which is deposited in the pond as sediment; in due course the storage capacity of the pond is abridged. Sedimentation control can be achieved through provision of grassed waterways, spillways, gully control measures, siltation traps *etc.*

b. *Seepage Control*

To arrest seepage losses of water, equal quantities of pond silt and clay are mixed and 0.2 per cent sodium carbonate is added to it.

c. *Evaporation Control*

The method for reducing evaporation loss has been to cover the water surface with a barrier that inhibits vaporization. On small tanks, a cover of roof is the obvious choice. But for bigger ponds the method comprises the use of liquid anti- evaporants like acetyl alcohol, floating bags and banana and palmyrah leaves that float on the water surface and reduce the area where vaporization occurs.

In brief various control measures are:

A. *Vegetative Measures (Agronomical Measures)*

1. Strip cropping
2. Pasture cropping
3. Grass land farming
4. Wood lands

B. *Engineering Measures (Structural Practices)*

1. Contour bunding
2. Terracing
3. Construction of earthern embankment
4. Construction of check dams
5. Construction of farm ponds
6. Construction of diversion
7. Gully controlling structure
   (a) Rock dam
   (b) Establishment of permanent grass and vegetation
8. Providing vegetative and stone barriers
9. Construction of silt tanks dentension

Influence of soil conservation measures and vegetation cover on erosion, runoff and nutrient loss.

Rainwater Harvesting

Rainwater harvesting is the main component of watershed management. Some of the watershed management structures are as follows:

- ☆ Broad beds and furrows
- ☆ Contour bunds
- ☆ Bench terraces
- ☆ Microcatchments
- ☆ Check dams
- ☆ Percolation ponds

Steps in Watershed Planning

- ☆ Preparation of base map for carrying out surveys.
- ☆ Details soil survey of the watershed for overall development.
- ☆ Assessing rainfall characteristics.
- ☆ Preparation of soil maps and classification for land for different uses according to capability for agriculture, forestry, horticulture *etc.*
- ☆ Preparation of inventory of existing land use and farm sizes.
- ☆ Appraisal of agricultural production patterns and potentials, present and potential markets and possible group action arrangements.
- ☆ Carrying out topographical and hydrological survey for engineering.
- ☆ Geo-hydrological survey to delineate areas suitable for ground water development.
- ☆ Formulation of an integrated time-bound plan for land and moisture conservation, ground water recharge, development of productive and productive afforestation, agriculture production, grass lands and horticulture.
- ☆ Assigning of priorities for the implementation of the project.
- ☆ Assessing social costs and benefits.

Tamil Nadu Watershed Development Agency (TAWDEVA)

- ☆ Integrated Watershed Management Programme (IWMP).
- ☆ Drought Prone Areas Programme (DPAP).
- ☆ Integrated Wasteland Development Programme (IWDP).
- ☆ National Watershed Development Project for Rainfed Areas (NWDPRA).
- ☆ Watershed Development Fund (WDF).
- ☆ Western Ghats Development Programme (WGDP).
- ☆ National Agriculture Development Programme (NADP).
- ☆ National Food Security Mission (NFSM).
- ☆ Agriculture Technology Management Agency (ATMA).

- ☆ Agriculture Resource Information Systems and Networking (AGRISNET).
- ☆ National Project on Management of Soil Health and Fertility.

Soil and Water Conservation Programmes

- ☆ Rain Water Harvesting and Run off Management Programme.
- ☆ Scheme for Artificial Recharge of Ground Water.
- ☆ Soil and Water Conservation in Tribal Areas under Integrated Tribal Development Programme.
- ☆ Hill Area Development Programme.
- ☆ Soil and Water Conservation in River Valley Project Catchments.

Water Management Programmes

- ☆ Command Area Development and Water Management Programme.
- ☆ World Bank Aided Tamil Nadu IAMWARM Project.

Recommendations for Best Irrigation Practices

The major agricultural use of water for irrigation and its supply is decreasing steadily due to competition with municipal and industrial sectors. Therefore, technological, managerial, policy innovation and human resources management are needed to increase the efficiency of use of the water that is available. Integrated water management in agriculture can be achieved by:

1. Reduction of Water Losses

Reduce of water lossesin the conveyance, distribution and application networks. Water leakages should be detected via advanced technologies, e.g. telemetry systems, GIS, remote sensing. Old water projects experiencing considerable water losses should be rehabilitated and modernized.

2. Improve the Efficiency of Irrigation System

Improvements in surface irrigation methods (efficiency 50-60 per cent) include land leveling and reduced widths and/or shorten lengths. For the on farm water distribution use gated pipes and lay flat pipes, buried pipes for basin and borders, lined–on farm distribution canals, good construction of on-farm earth canals, easier control of discharges and control of seepage. Improvements in sprinkler irrigation systems (efficiency up to 85 per cent) include the adoption or correction of sprinkler spacing, the design for pressure variation not exceeding 20 per cent of the average sprinkler pressure, the use of pressure regulators in sloping fields, the monitoring andadjustment pressure equipment, application of irrigation during no windy periods, adoption of smaller spacing and large sprinkler drops and application rates in windy areas, the adoption of application rates smaller than the infiltration rate of the soil and careful system maintenance. Improvements in localized irrigation systems (Efficiency 95 per cent) aiming to reduce the volumes of water applied and increase the water productivity include the use of a single drip line for a double row crop, the use of micro-sprayers in high infiltration soils, the adjustment of duration of water application and timing to soil and crop characteristics, the control of pressure and discharge variations, the use of appropriate filters to the water quality and the emitter characteristics used, the adoption of careful maintenance and automation.

3. Increase Water Use Efficiency

Increase water use efficiencyCan be achieved with the obligatory use of localized irrigation systems by the farmers (with or without subsidies), the proper irrigation scheduling according to actual needs of the crops, the establishment of a system for advising farmers on their irrigation schedules, the introduction of appropriate agronomical practices and the application of salinity management techniques.

4. Adoption of Innovative Irrigation Techniques

In water scarce regions irrigation approaches not necessarily based on full crop water requirements like regulated deficit irrigation (RDI) or subsurface irrigation (SSI) must be adopted. Fertigation (efficient fertilizer application) and chemigation (easy control of weeds and soil born diseases) should also be promoted.

5. Water Pricing

For proper water pricing volumetric water metering and accounting procedures are recommended. Progressive, seasonal and over-consumption water tariffs as well as temporary drought surcharges rates contribute to water savings and should be promoted. Furthermore, an increasing block tariff charging system, that discourages water use levels exceeding crop's critical water requirements, must be established. It will be the basis for promoting conservation, reducing losses and mobilizing resources. Furthermore, it could affect cropping patterns, income distribution, efficiency of water management, and generation of additional revenue, which could be used to operate and maintain water projects.

6. Reuse of Marginal Waters (reclaimed or brackish) for irrigation

Reclaimed waters can be used under some restriction for irrigation of tree, row and fodder crops. In addition to water they provide the soil with nutrients, minimizing the inorganic fertilizer application. Treated sewage is looked upon with skepticism by farmers. They instead prefer to use surface and/or groundwater. Special effort should be given in educating farmers to accept treated sewage. In addition the tariff for this source of water should be lower than the tariff of the primary sources. This may not be difficult to achieve because the primary and secondary levels of treatment are regarded as sunk costs since they are required by the new WFD. When using low quality water for example brackish or saline water an integrating approach for water, crop (salt tolerant varieties) field management (suitable tillage) and irrigation system (adequate leaching, suitable devices) should be considered.

7. Wider and more Effective Participation

Wider and more effective participation of the public, NGOs and end users for the preparation of the plans, in decision-making, in monitoring the implementation and generally in the management of water. The participation of these groups in the above processes safeguards the acceptability of the plans by the general public, raises support on the part of the body politics and promotes success in possible conflict resolutions.

8. Capacity building

The existing "capacity building" is poor. It requires an appropriate mix of competent personnel, technologically advanced devices and facilities, legal guidelines and administrative efficient and effective processes for the integrated management of the water resources.

a. Education and Training

Education and trainingof professional, technical staff and decision makers and others, including non-public organizations, on a wide range of subjects related to integrated water management.

b. Manpower Build Up

Institutions to be staffed with qualified manpower (managers, engineers, technicians, social scientists) that should be adequately compensated.

c. Facilities and Procedures

Water authorities at all levels of management should be equipped with technologically advanced devices and programs e.g. computers and software for the application of new techniques such as GIS, remote sensing *etc.* These advanced techniques facilitate the multi-sectoral information availability and use and help water managers in their decision- making.

d. Legislative Changes

The fragmented and antiquated legislation should be promoted. Responsibility of water resources planning and operation, especially at the overall national level, should be under one institution for proper organization and non-conflicting actions by the various authorities. Water authorities should participate fully in the formulation of agricultural policies because the development of water and land should be fully integrated. In practice, agricultural decisions are water decisions and vice versa.

References

agritech.tnau.ac.in/./agri_majorareas_**watershed_watershed**mgt.html

Balusamy M. 2002. Land use planning and watershed management in rainfed agriculture. Centre of advanced studies department of agronomy TNAU, coimbature.

Chartzoulakis K. 2012. Sustsainable Water Management in Agriculture under Climate Change. NAGREF, Institute for Olives and Subtropical Plants, Greece. en.wikipedia.org/wiki/**Watershed_management** FAO, 1999. The state of food insecurity in the world. Rome, Italy.

Rosegrant M.W. and Cai X., 2002. Global water demand and supply projections: Results and prospects to 2025. Water Intern., **27:** 170-182. rural.nic.in/sites/IWMP.asp

www.dardni.gov.uk.

Chapter 3

Instant Messaging Tool: An Effective Connecting Platform among Stakeholders

M. Senthil Kumar[1], H. Philip[2] and P. Sivaraj[3]

[1]Assistant Professor (Ag. Extn.)
[2]Director of Extension Education and [3]Ph.D., Scholar
Tamil Nadu Agricultural University, Coimbatore – 641 003, Tamil Nadu

---

Information and Communication Technology is ruling the world in all walks of life and access to mobile phones and internet facility is growing in India at a rapid rate in recent years. However, the access to internet based technologies have confined primarily to the urban areas. Rural communities have not been able to gain to the same extent from IT. As a means of agricultural technology transfer to farmers, information technology has had a limited impact.

Over the years, Radio and TV have vastly increased their reach, as also reception facilities and played a major role in communicating with an audience with low literacy skills. The traditional electronic media played a key role in rapid and effective dissemination of general information and advice to farming communities. This includes market information; market led production planning, on farm and posts harvest management/value addition, e-contracting, market networks and market intelligence. The World Wide Web especially portals, web sites and other pull technological services are also contributing to the technology transfer of general information/knowledge.

Communication technologies that enable or facilitate user-to-user interactivity and interactivity between user and information. *i.e.* "one-to-many" model of traditional mass communication with the possibility of a "many-to-many" web of communication.

New media: Consists of a number of technologies that facilitate interactions among stakeholders using a variety of web- or mobile-based tools and technologies. "Web 2.0" and "social media" are umbrella terms that encompass the various activities that utilize digital technologies, social engagement, and content delivery. Such activities involve many technologies and communication methodologies including, but not limited to, blogs, photo and video sharing services, social networking, and geospatial mapping tools, discussion forums, and wikis. These technologies may enable social tagging and bookmarking and mobile messaging.

Scope for Using Instant Messaging Tools in India

Although the Internet penetration in India is very low at 19 per cent, India has the third largest Internet user base in the world out of which more than 50 per cent are mobile-only internet users.

In India, the number of people who own mobile phones is greater than the number who own personal computers. The Indian government is committed to setting up a robust digital infrastructure and to promote adoption of mobile Internet and related products and services. In 2014-15, the Government budgeted INR 500 crore for building infrastructure as per the National Rural Internet and Technology Mission with an additional INR 100 crore budgeted for improving e-governance.

India is making the transition from features phones to smartphones rapidly. This is accelerated by the availability of low-cost smartphones and data plans. SMS, email, messaging and social networking apps are the most popular used apps, while video streaming and banking services are the least used apps.

Instant Messaging tools

The most prominent Instant messaging tools are WeChat (468 million active users), Viber (209 million active users) and LINE (170 million active users) and WhatsApp (900 million users).

Why Instant Messaging Tool for Agricultural Extension?

- ✰ To combat the demand for technology transfer from Globalization and Modernization
- ✰ To tap the challenges and opportunities across the world
- ✰ Sharing of knowledge across the world
- ✰ Public Expenditure in Extension could be reduced
- ✰ Effective utilization of available platforms
- ✰ Penetration of smart phones even in villages
- ✰ Accessibility of connectivity

WhatsApp as a Tool for Sharing of Agricultural Knowledge

WhatsApp Messenger is a cross-platform mobile messaging app and it is a platform to share real-time information which allows user to exchange messages, audio, video, photographs. It allows users to create groups, broadcast/send unlimited images, video and audio media messages simultaneously to one person, or to a group. WhatsApp Messenger requires a smart mobile phone with internet data access.WhatsApp is the largest community of users of any IM client on any device, ever, has been on the rise since its inception. With over 600 million users it is the most popular application for Instant Messaging, and has found mainstream acceptance and popularity now world over.As of September 2015, WhatsApp had a user base of up to 900 million, making it the most globally popular messaging application.

Steps in using WhatsApp as a tool for Effective and efficient sharing of Agricultural Technology to select farmers.

- ✰ Scientists and farmers have to be trained in the use of Instant Messaging application namely WhatsApp in creating, sharing, recording and retrieval of agricultural information/technology.
- ✰ Access to be made to necessary hardware and connectivity namely availability of Smart Mobile devices and Internet connectivity to all the stakeholders (Farmers/Scientists) involved in the project. The knowledge sharing process has to be documented by all the stakeholders.
- ✰ The reusability of shared knowledge retrieved from this extension model by for research and extension may be studied and the constraints faced by the users in using WhatsApphas to be documented.

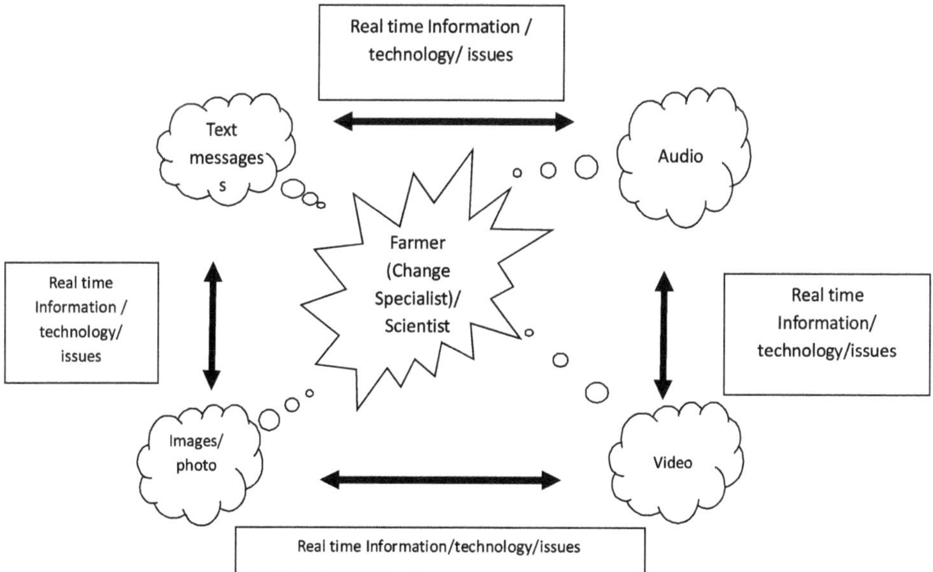

Figure 3.1: Model Depicting the Technology Sharing Model through WhatsApp.

1. Success Stories on Using WhatsApp for Sharing of Knowledge Platform from Punjab

The group, 'Young Innovative Farmers', was set up by Gurdaspur Agriculture Development Officer Dr.Amrik Singh on August 15, 2014 has over 100 members, including 90 farmers and 10 experts. Moreover, the farmers in the group have set up their own groups with local farmers to disseminate the information,"related to crop health, seed procurement, soil health, use of fertilisers and pesticides and now the farmers in Punjab can now get immediate advice via a WhatsApp group which includes agricultural experts.

A top official of Punjab said that "We have a shortage of staff in our department. If we visit a few villages daily, we can only meet a few dozen farmers. But on WhatsApp we can interact with several farmers daily," The farmers also upload photographs of their disease-hit crops to seek advice and solutions are provided back to him.

2. Success Story of using WhatsApp for Sharing of Knowledge Platform from Belagavi, Karnataka

Horticultural Producers' Co-operative Marketing was started in 1959 to ensure Karnataka farmers were able to market their produce directly to consumers, thus eliminating the middlemen and the vegetables and fruits are purchased directly from the farmers at market price.Three HOPCOMS stores and consumers have formed a group on WhatsApp where the store owners send details of newly arrived produce along with images to the interested customers of the WhatsApp group.If the customer is dissatisfied or the produce is damaged, he can upload images and get the items replaced. It was noticed by HOPCOMS that more number of people are visiting the store.

3. Success Story of using WhatsApp for Sharing of Knowledge Platform from Belagavi, Karnataka

The Commissioner of the Agriculture Department of karnatakastarted a platform where the officials and farmers could interact and so created a WhatsApp group with farmers and few officials to discuss the agricultural issues encountered by the farmers and resolve them. One group called "Progressive Farmers" has 198 members who discuss advance farming techniques.A farmer belonging to one region raises his issue and uploads photographs and a farmer elsewhere responds with an appropriate solution.

VeenaSudhir, who practices integrated livestock farming in Madikeri seconds Yusuf's opinion. "Such groups help is clearing the doubts and sharing the ideologies," she said. Citing an instance of how farmers came to her rescue, "Weed had grown abundantly and posed a threat to my farm. Having tried different weedicides, it did not alleviate the problem. I sought the help of the farmers on the group and they suggested a different brand of weedicide which worked," she explained.

4. Initiative by a Rajkot-Based Agriculture Entrepreneur

A Rajkot-based agriculture entrepreneur Dinesh Tilva has turned popular social media app WhatsApp into a classifieds marketplace that allows farmers to trade goods such as grains, vegetables, seeds, irrigation equipment and tractors, among others. He moderates all the ads and broadcasts them to nearly 1,500 contacts in his smartphone. Most of these contacts are farmers, while some are traders of fertilizers, seeds, farm equipment, while a few are agents dealing in land and property. Farmers from across Gujarat, mostly from Saurashtra and central parts, find the system beneficial for them. "This practice is very convenient for us. If we wanted to buy a Gir-breed cow, it usually takes an entire day to locate one and if it isn't fit, you just waste the day. But now, we can see the picture of the cattle and decide accordingly. It saves time, energy and money," says Jagani of Kuvadva village. Tilva says most farmers own entry-level smartphones which are compatible with WhatsApp. "Those who are unable to participate in the system are assisted by their peer or relatives,"

5. Farmer's Friend Group by Extension Machinery of Bangalore

The entire department of Agriculture in Bangalore going on WhatsApp in a new group called Farmer's Friend.The officials of the department uses WhatsApp for issuing circulars, news, policy matters best practices like harvesting, farming and crop cutting are shared instantly with photographs and videos in the free platform that connects farmers from different parts of state through the department field staff.

The commissioner, director, additional director and director and joint directors and assistant directors are all in the newly created group. The initiative helps at two levels - effective monitoring of staff and connecting with farmers.

"Gone are the days of one-sided information dissemination. By going on WhatsApp, we are creating a platform to share best practices and take the state government schemes to them as soon as they come into existence," Subodh Yadav, the state agriculture commissioner, told Bangalore Mirror.They are also creating Facebook pages and an interactive site wherein queries of the 78 lakh farming families are addressed instantly by experts. "Information reaches us instantly as the documents are flashed on WhatsApp and the same day it is taken to farmers," a senior officer pointed out.

6. WhatsApp Group by KVKs of TNAU

The Directorate of Extension Education, Tamil Nadu Agricultural University, Coimbatore has started a WhatsApp group in 2015 to facilitate scientists from TNAU KVKs to share information, knowledge on agricultural technologies, to share their activities, field visits, training programmes, success stories among the scientists.

The Director of Extension Education, Programme Coordinators and Scientists and staffs of TNAU are members of the group named KVKs TNAU. The group has 85 active members as of October 2015.

The group shares valuable information on programmes, campaigns, field visits, farmers success stories and press news regularly. The information helps in sharing of knowledge instantly.

Advantages of using Instant Messaging Tools

- Easy to learn and do
- Students can easily create forums, and share
- Peer review possible
- Easy Storage and retrieval
- Reproducibility
- Audio visual learning

Students

- Generate, collate, Store and retrieve and discuss reading materials
- Document case studies in audio and video format
- Network with millions of likeminded colleagues
- Improves interaction

Teachers

- Easy to use and address the concerns of millions of students
- Teaching at own phase
- Effective communication through the use of audio- visual aids
- Improvisation over time
- Create learners interest

Researchers

- Open and real time research information made available
- Can reach and access information across the world
- Reduces plagiarism
- Improves the research quality
- Can organize or group, researchers of same interest across the world

Extension

- Up to date advisory for farmers, extension workers
- Facilitates easy understanding/diagnosis of problems and providing quality solution.
- Helps to create, store and retrieve local database of farmers, problems, solutions

Conclusion

- Proactive use of Instant Messaging tools will bring a knowledge revolution in agriculture

- ✰ Extension scientists need to capitalize the use of modern instant messaging tools for the benefit of the farming community, research scholars and future generation.

References

Source: http: //yourstory.com/2015/07/mobile-internet-report-2015/

https: //en.wikipedia.org/wiki/WhatsApp

http: //indianexpress.com/article/india/punjab-and-haryana/punjab-farmers-turn-to-whatsapp-group-for-farming-solutions/#sthash.OMHv07NC.dpuf

http: //www.thenewsminute.com/article/how-farmers-karnataka-are-using-whatsapp-connect-consumers#sthash.fYEGbs0b.dpuf

http: //www.thehindubusinessline.com/economy/agri-business/whatsapp-turns-a-trading-platform-for-gujarat-farmers/article7154940.ece

http: //www.bangaloremirror.com/bangalore/others/Agriculture-dept-top-brass-goes-WhatsApping-with-farmers/articleshow/29827557.cms

vi. Extension scientists need to popularize the use of modern instant messaging tools for the benefit of the farming community, research scholars and future generation.

References

Source: http://vyolusiony.com/2015/07/mobile-internet-report-2015/

https://en.wikipedia.org/wiki/WhatsApp

http://indianexpress.com/article/india/punjab-and-haryana/punjab-farmers-turn-to-whatsapp-group-for-farming-solutions/#sthash.ONfTyhNC.dpuf

[illegible]

[illegible]

Chapter 4

Integrated Crop Management for Better Agribusiness

C. Jayanthi, M. Balusamy, M. Ananthi and P. Gnanasoundari

Directorate of Crop Management, Tamil Nadu Agricultural University, Coimbatore – 641 003, Tamil Nadu

Integrated Crop Management is particularly appropriate for small farmers because it aims to minimize dependence on purchased inputs and to make the fullest possible use of indigenous technical knowledge and land use practices. Integrated crop management is a method of farming that balances the requirements of running profitable business with responsibility and sensitivity to the environment. It includes practices that avoid waste, enhance energy efficiency and minimize pollution. Integrated crop management combines the best of modern technology with some basic principles of good farming practice. Integrated crop management is a whole farm, long term strategy. It cannot be applied to one crop, or one field, or one season. Although primarily concerned with crop production, livestock management is equally important on mixed farms because livestock are consumers of crops and providers of organic nutrients.

Key Components of Crop Management

- Land management
- Integrated nutrient management
- Crop rotation

- ✰ Integrated weed management
- ✰ Water management
- ✰ Organic farming
- ✰ Integrated farming system

Land Management

Implements that can carryout tillage and land shaping in one single operation will help in saving time and cost. If land preparation, land shaping and sowing can be done in one single operation it can save considerable time. This is termed as once over tillage, plough planting or conservation tillage. Suitable tractor drawn machinery like a broad bed former cum seeder, basin lister cum seeder which can complete the land shaping and sowing simultaneously can be used.

i) Minimum/Reduced Tillage

The objectives of these systems include (a) reducing energy input and labour requirement for crop production (b) conserving soil moisture and reducing erosion (c) providing optimum seedbed rather than homogenizing the entire soil surface, and (d) keeping field compaction to minimum. This is the method of preparation of land or seed bed with minimum disturbance of soil.

ii) Conservation Tillage

The objectives are to achieve soil and water conservation and energy conservation through reduced tillage operations. Both systems usually leave crop residue on the surface and each operation is planned to maintain continuous soil coverage by residue or growing plants. The conservation tillage practices may advance some of the goals of alternative farming such as increasing organic matter in soil and reducing soil erosion, but some conservation tillage practices may increase the need for pesticides. Conservation tillage changes soil properties in ways that affect plant growth, and reduce runoff from fields. The mulched soil is cooler and soil surface under the residue is moist, as a result many conservation tillage systems have been successful. Here tillage operations are restricted either by cutting down its number or by carrying out as many operations as possible in one pass and allow atleast 30 per cent of residues remain on the soil surface. New studies have been undertaken in recent years to examine the effects of different type of conservation tillage on soil erosion rates, soil conditions and crop yields. The results show that the success is soil specific. It also depends upon how weeds, pests and diseases have been controlled. Generally, the better-drained, coarse and medium-textured soils with low organic matter content respond best. The systems are not successful on poorly drained soils with high organic contents or on heavy soils, where use of the mould board plough is essential.

iii) Zero Tillage/No-till System

The practice of no tillage, whereby drilling of seeds takes place directly into the stubble of the previous crop and weeds are controlled by herbicide, has been found to increase the percentage of water stable aggregates in the soil compared with tyne or disc cultivation and ploughing (Aina, 1979; Doglus and Goss, 1982). In this type of tillage all crop residues are maintained on the soil surface, which reduce erosion and enhances water conservation. For no tillage to be effective, adequate plant residue must be available. No-tillage cannot be effective in degraded soils with low plant residue. Where residue ranges from 1.5 to 2.5 tonnes/ha and weeds are controlled by herbicides, water storage and crop yields are similar to fields with stubble mulch tillage.

iv) Strip-Zone/Skip Tillage

In this system the tillage operations are concentrated only on the plant rows where plants grow leaving the inter-row areas untilled allowing the crop residue to be deposited. Depending upon the crop to be grown and the land situation its planning may be designed.

v) Mulch Tillage

In this system large percentage of plant residue (leaf, stubble, stalk, crown, root) is left on or near the surface as protective cover. Mulch tillage is accomplished with implements (sweeps or blades) that undercut the soil surface, thereby retaining most of the crop or plant residue on the surface. One of the main functions of this system is to break or disturb the continuous capillaries through which the evaporation loss of soil moisture from root zone takes place. Mulch tillage can also be given with a chisel plough, but rotary rod weeder may be needed to improve the weed control. The problem of weed control occurs when mulch tillage is performed in moist soils or when precipitation occurs soon after tillage.

Integrated Nutrient Management

"The primary goal of integrated nutrient management (INM) is to combine old and new methods of nutrient management into ecologically sound and economically viable farming systems that utilize available organic and inorganic sources of nutrients in a judicious and efficient way. Integrated nutrient management optimizes all aspects of nutrient cycling. It attempts to achieve right nutrient cycling with synchrony between nutrient demand by the crop and nutrient release in the soil, while minimizing losses through leaching, runoff, volatilization and immobilization".

Concepts of INM

The concept of Integrated Nutrient Management (INM) takes into consideration the nutrient cycle involving soils, crops and livestock, nutrient deficiencies, organic recycling, conjunctive use of organic manures and mineral fertilizers and biological nitrogen fixing potential. Further, the basic concept of Integrated Nutrient Management (INM) is the adjustment of soil fertility and plant nutrient supply at an optimum level for sustaining the desired crop productivity. Absolute and relative application of nutrients is a part of INM.

Components of INM

Fertilizers

Fertilizers continued to be the most important ingredient of INM. The dependence on fertilizers has been increasing constantly because of the need to supply large amounts of nutrients in intensive cropping with high productivity.

Organic Manures

Organic manures like urban compost, FYM, crop residues, human excreta, city refuse, rural compost, sewage sludge, pressmud and other agro industrial wastes have large nutrient potential. Compost and FYM have traditionally been the important manures for maintaining the soil fertility and ensuring yield stability. Other potential organic sources of nutrients are non edible oil cakes and wastes from food processing industry. These manures, besides supplying nutrients to the first crop, also leave substantial residual effect on succeeding crops in the system.

Legumes

Legumes have a long standing history of being soil fertility restorers due to their ability to obtain N from the atmosphere in symbiosis with rhizobia. Legumes could prove an important ingredient of INM when grown for grain or fodder in a cropping system, or when introduced for green manuring.

i) Legumes grown in Rotation

Different legumes have the capacity to leave different amounts of N for use by the succeeding crop. The carryover of N for succeeding cereal crop is 60-120 kg in berseem,75 kg in Indian clover, 35 - 60 kg in fodder cowpea, 55 kg in blackgram, 60 kg in groundnut, 68 kg in gram and 50 kg in lathyurs per ha. Grain yield of succeeding crop increased with preceding legume than a preceding cereal crop. In general, pulse crops leave a residual amount of 30 to 50 kg N/ha for utilization by the succeeding crop.

ii) Legumes as Green Manures

The decomposing green manure has a solubilising effect on N, P, K and microorganisms in the soil. It also reduces leaching and gaseous losses of N. Besides it improves the physical, chemical and biological properties of soil.

Crop Residues

Crop residues have several competitive uses and may not be always available as an ingredient of INM, yet where mechanical harvesting is practiced, a sizeable quantity of residues is left in the field, which can form a part of nutrient supply.

Biofertilizers

Biofertilizers are the materials containing living or latent cells of agriculturally beneficial microorganisms that play an important role in improving soil fertility and crop productivity due to their capability to fix atmospheric N, solublize/mobilize P and decompose farm waste in the release of plant nutrients.

Crop rotation is one of the oldest and most effective cultural strategies. It means the planned order of specific crops planted on the same field. It also means that the succeeding crop belongs to a different family than the previous one. The planned rotation may vary from 2 or 3 years or longer period. Some insect pests and disease-causing organisms are hosts specific. For example, rice stem borer feeds mostly on rice. If we don't rotate rice with other crops belonging to a different family, the problem continues as food is always available to the pest. However, if we plant legume as the next crop, then corn, then beans, then bulbs, the insect pest will likely die due to absence of food.

Advantages of Crop Rotation

- Prevents soil depletion
- Maintains soil fertility, soil organic matter levels and soil structure
- Reduces soil erosion
- Controls insect pests.
- Reduces reliance on synthetic chemicals
- Reduces the pests' build-up
- Prevents diseases
- Helps control weeds

Integrated Weed Management

The cultivation of high yielding crop varieties responsive to fertilizers and irrigation and the new intensive cropping systems have brought to the forefront the problem of weeds which cause tremendous losses to crops and their produce. IWM is about NOT relying on only one or two methods of weed control alone, and particularly not relying on herbicides alone. An IWM program uses a range of methods of weed control in combination so that all weeds are controlled by atleast one component of the weed management system. Ultimately, the aim of IWM is to prevent weeds setting seeds, or vegetatively reproducing, so that the weed population is reduced over time, reducing weed competition and improving crop productivity. Weed management approaches that rely on a limited number of strategies often end up with uncontrolled weeds. The most common example of this is the repeated reliance on one or two groups of herbicides to control a target weed population. Within a weed population there is likely to be individual plants that are naturally resistant to any single herbicide. The frequency of these resistant individuals in the population is usually very low. Repeated exposure of the weed population to a limited range of herbicides results in these resistant individuals being selected out, so that eventually a large proportion of the population is resistant to the herbicides. Eventually herbicide resistance develops such that the herbicide no longer controls the target weed.

As well as selecting for herbicide resistant weeds, the repeated use of a small number of weed management tools causes a species shift in the weed population. Weed species that are not controlled by these management tools will soon dominate the weed population, and the weed spectrum will shift towards these weeds. This species shift can result in new weed problems, with weed species that are much more difficult to control than were the original weeds. The risk of developing these problems can be greatly be reduced by using an IWM program. All the individual components of the system contribute to a total weed management system.

Why to adopt IWM?

Using an IWM program throughout the entire crop rotation, including rotation of crops and fallows, will:

- ☆ Reduce the reliance on herbicides,
- ☆ Reduce the risk of herbicide resistance developing in the weed spectrum and prolong the usefulness of the available herbicides,
- ☆ Reduce the rate of shift in the weed spectrum towards more herbicide tolerant weeds,
- ☆ Reduce the risk of herbicides accumulating in the soil and riverine systems, and
- ☆ Reduce the total weed control costs in the future by reducing the weed seed bank (the number of weed seeds in the soil).

Need (IWM)

One method of weed control may be effective and economical in a situation and it may not be so in other situation.

1. No single herbicide is effective in controlling wide range of weed flora.
2. Continuous use same herbicide creates resistance in escaped weed flora or causes shift in weed flora.
3. Continuous use of only one practice may result in some undesirable effects (*e.g.*) Rice – Wheat cropping system – Phalaris minor).
4. Single method of weed control may lead to increase in the density of particular weed.
5. Indiscriminate herbicide use and its effect on the environment and human health. IWM is the rational use of direct or indirect control methods to provide cost effective weed control. Such an approach is the most attractive alternative from agronomic, economic and ecological point of view.

IWM Approach

Development of alternative weed control methods for reduced tillage system will continue to be important because tillage or hand weeding still is the major practicable alternative to weed control with herbicides. Alternative reduced or non-chemical methods for intercropping system may be of special interest in tropical and under developed countries where herbicides for these systems are either unavailable or unaffordable, yet an alternative to land cultivation is desired. With sound crop-weed management practices and integration of new cultural practices and herbicide technologies with traditional preventive practices both acceptable levels of weed control and improved environmental quality can be achieved.

Integrated Crop Management Practices for Important Crops

Rice

Transplanted Rice

Plough the land during summer to economize the water requirement for initial preparation of land. Flood the field 1 or 2 days before ploughing and allow water to soak in. Keep water to a depth of 2.5 cm at the time of puddling. Apply 500 kg of gypsum/ha at last ploughing.Optimum age of the seedlings is 21 days for short duration varieties with 15 x 10 cm spacing, 28 days for medium duration varieties with 20 x10 cm spacing and 35 days for long duration varieties with spacing of 20 x 15 cm. Apply 12.5 t of FYM or compost or green leaf manure @ 6.25 t/ha and 150:50:50 kg of NPK/ha. Apply N in four equal splits *viz.*, basal, tillering, panicle initiation and heading stages and P 100 per cent as basal and incorporation. K in two splits (Basal and PI). Micronutrient mixture 12.5 kg/ha or 25 kg of zinc sulphate enriched in 250 kg FYM. For foliar application, 1 per cent urea + 2 per cent DAP – 1 per cent KCl at PI and 10 days later for all varieties for yield maximization is recommended. For water management, 2 cm of water up to seven days after transplanting and increase the depth of irrigation to 2 to 5 cm progressively with the crop stage. Use pretilachlor 0.75 kg ha^{-1} on 3 DAS + 2,4 DEE 0.4 kg ha^{-1} on 15 DAT (or) Pre emergence application of butachlor 0.75 kg/ha + bensulfuron methyl 50 g/ha on 3 DAT followed by mechanical weeding on 45 DAT is effective for broad spectrum weed control.

Wet Seeded Rice

Wet seeded rice can be followed in all the areas wherein transplanting is in vogue. Sow the pre-germinated seeds by drum seeder along with green manure with thin film of water. The seed raterecommended is 60 kg/ha. Thinning and gap filling is to be done at 14 – 21 DAS. Incorporated the green manure at 30 DAS using cono-weeder. For nutrient management, application of 12.5 t of FYM or compost or green leaf manure @ 6.25 t/ha and 150:50:50 kg of NPK/ha is recommended. Application of N in four equal splits *viz.*, basal, tillering, panicle initiation and heading stages and for P as basal and incorporated in soil. For foliar application, 1 per cent urea + 2 per cent DAP – 1 per cent KCl at PI and 10 days later for all varieties for yield maximization is recommended.

For water management, wet the soil by thin film of water for first one week of crop growth. Increase the depth of irrigation to 2.5 cm progressively with the crop stage. Use pretilachlor 0.45 kg ha^{-1} on 3 DAS + roto cylindrical weeder weeding on 45 DAS in wet seeded rice resulted in excellent control of weeds.

Rainfed Rice

Dry plough to get fine tilth taking advantage of rains and soil moisture availability and apply gypsum at 1 t/ha basally wherever soil crusting and soil hardening problem exist. Seed rate of 75kg/ha dry seed is recommended. Seed hardening with 1 per cent KCl for 16 hours (seed and KCl solution 1:1) and shade dried to bring to storable moisture. On the day of sowing, treat the hardened seeds first with *Pseudomonas fluorescens* 10g/kg of seed and then with *Azophos* one kg or *Azospirillum and Phosphobacteria* @ one kg each per ha of seed. The blanket recommendation of fertilizer is 50:25:25 kg N: P: K/ha.Apply a basal dose of 750 kg of FYM enriched with fertilizer phosphorus (P at 25 kg/ha). Apply N and K in two equal splits at 20 - 25 and 40 - 45 days after germination. Foliar spray of 1 per cent urea + 2 per cent DAP + 1 per cent KCl at PI and 10 days later may be taken up for enhancing the rice yield. For weed management, apply pendimethalin 1.0 kg/ha on 5 days after sowing or pretilachlor + safener (Sofit) 0.45 kg/ha on the day of receipt of soaking rain followed by one hand weeding on 30 to 35 days after sowing.

Rainfed Sorghum

For land preparation, plough with Chisel/disc plough once in 3 years with broad bed furrow (120 x 30 cm).The recommended seed rate is 15 kg/ha. The seeds are pre-soaked in 2 per cent potassium dihydrogen phosphate solution for 6 hours in equal volume and sowing with seed drill is practiced with the spacing of 45 x15 cm. Apply 12.5 t/ha of Composted Coir pith + NPK at 40:20:20 kg/ha for nutrient management. For control of weeds, pre – emergence application of Atrazine 0.25 kg/ha (3 DAS) + one hand weeding (25-30 DAS) is to be practiced.

Rainfed Finger Millet

For preparation of land, ploughing with chisel/disc plough (once in 3 years) and form compartmental bunding. 5 kg/ha of seeds is to be used. Seeds are to be treated with *Pseudomonas fluorescence* 10 g/kg or *Trichoderma viride* @ 4g/kg seed + Azophos (40 g/kg of seed) and should be broadcasted. The spacing adopted is 22.5 X10 cm. Recommended dose of fertilizer of NPK is 40:20:20 kg/ha is to be adopted. For effective weed control, application of Isoproturon 0.5 kg/ha and Inter cultivation on 25-30 DAS.

Irrigated Redgram

For preparation of land, ploughing with chisel/disc plough (once in 3 years) and forming compartmental bunding is recommended. For raising the pure crop, the seed rate for varieties Co 6,VBN 2 (10 kg/ha), VBN 1, VBN (Rg) 3,APK 1 (20 kg/ha), CO 5,COPH2,CO (Rg) 7 (25 kg/ha) and for intercrop VBN 2,CO 6, COH1 (5 kg/ha), CO 5 (12.5 kg/ha), VBN 1, APK 1 (10 kg/ha) and for SA 1,BSR 1 (50 g/100 m length bund). For seed treatment, treat with *Pseudomonas fluorescence* 10 g/kg or *Trichoderma viride* @ 4 g/kg seed + Rhizobium and Phosphobacteria @ 600 g per ha. The recommended NPKS: 25:50:25:20 kg/ha + $ZnSO_4$ 25 kg/ha (basal) 50 per cent nitrogen can be substituted through organic source (850 kg of vermicompost per hectare). *Rhizobium* and Phosphobacteria @ 2 kg per ha recommended as soil application. Water should be provided at sowing, 3 DAS, bud initiation, 50 per cent flowering and pod development stages. For effective weed control, application of pendimethalin 1 kg/ha + one hand weeding (25 – 30 DAS).

Rainfed Redgram

For land preparation, ploughing with chisel/disc plough (once in 3 years) followed by cultivator is recommended. For raising crop, the seed rate for varieties

Co 6,VBN 2 (10 kg/ha), VBN 1, VBN (Rg) 3,APK 1 (20 kg/ha), CO 5,COPH2,CO (Rg) 7 (25 kg/ha) is recommended. Treat the seeds with *Pseudomonas fluorescence* 10 g/kg or *Trichoderma viride* @ 4 kg seed followed by *Rhizobium* and Phosphobacteria @ 600 g per ha and sowing using seed drill. The requirement of fertilizer (NPKS) is 12.5:25:12.5: 10 kg/ha + $ZnSO_4$ 12.5 kg/ha (basal) 100 ppm salicylic acid at first flowering and 15 days after flowering. For weed control, apply pendimethalin 1 kg/ha + one hand weeding (25 – 30 DAS).

Irrigated Blackgram

Plough the field with 2-3 times to fine tilth. For raising the pure crop, the seed rate for varieties T9, Co 5,ADT 5,APK 1, VBN (Bg) 3, VBN (Bg) 4 (20 kg/ha), ADT 3,ADT 4 (Rice fallows) (25 kg/ha), CO 4, ADT 2 (bunds) (50 g/100 m length) and for mixed crop T9,CO 5,VBN 2,VBN (Bg) 3, VBN (Bg) 4 (10 kg/ha) with the spacing of 30 x10 cm. For seed treatment, treat with *Pseudomonas fluorescence* 10 g/kg or *Trichoderma viride* @ 4 g/kg seed + Rhizobium and Phosphobacteria @ 600 g per ha and micronutrients (Zn,Mo and Co @4,1,0.5 g/kg of seed). The fertilizer dose of NPKS: 25:50:25:20 kg/ha + $ZnSO_4$ 25 kg/ha (basal) 50 per cent nitrogen can be substituted through organic source (850 kg of Vermicompost per hectare). The critical stages of water requirement are sowing, 3 DAS and at 10-15 days interval (bud initiation, 50 per cent flowering and pod development stages). Application of pendimethalin 1 kg/ha + one hand weeding (25 – 30 DAS) is effective for weed control.

Rainfed Blackgram

Plough the field with 2-3 times to fine tilth. For raising the pure crop, the seed rate for varieties T9, Co 5,ADT 5,APK 1, VBN (Bg) 3, VBN (Bg) 4 (20 kg/ha), and for mixed crop T9, CO 5,VBN 2,VBN (Bg) 3, VBN (Bg) 4 (10 kg/ha) with the spacing of 30 X10 cm. For seed treatment, treat with *Pseudomonas fluorescence* 10 g/kg or *Trichoderma viride* @ 4 g/kg seed + *Rhizobium* and Phosphobacteria @ 600 g per ha and micronutrients (Zn, Mo and Co @4,1,0.5 g/kg of seed) using seed drill for sowing. The fertilizer dose of NPKS: 12.5:25:12.5:10 kg/ha + $ZnSO_4$ 25 kg/ha (basal). Application of pendimethalin 1 kg/ha + one hand weeding (25 – 30 DAS) is effective for weed control.

Irrigated Groundnut

For preparing land, one disc ploughing and ploughing with cultivator (twice) is to be done. Use 120 kg/ha of kernels and adopt the spacing of 30 x 10 cm. Treat the seeds with *Trichoderma viride* @ 4 g/kg of seeds or *Pseudomonas fluorescence* 10 g/kg of seeds and seed hardening with 0.5 per cent $CaCl_2$ for 6 hrs. Sowing is line sowing with seed drill at 5 cm depth. The fertilizer requirement of NPK is 17: 34: 54 kg/ha as basal and gypsum @ 400 kg/ha (200 kg at basal and 200 kg at second hand weeding followed by earthing up. Irrigation should be done at sowing, life irrigation on 7th DAS, pre flowering irrigation on 25th DAS, at flowering and pod filling stages. For effective weed management, Pre – emergence application of pendimethalin 0.75 kg/ha followed by hand weeding at 45 DAS.

Rainfed Groundnut

For preparing land, one disc ploughing and ploughing with cultivator (twice) is to be done. Use 140 kg/ha of kernels and adopt the spacing of 30 X10 cm. Treat the seeds with *Trichoderma viride* @ 4 g/kg of seeds or *Pseudomonas fluorescence* 10 g/kg of seeds and seed hardening with 0.5 per cent $CaCl_2$ for 6 hrs. Sowing is line sowing with seed drill at 5 cm depth. The fertilizer requirement of NPK is 10: 10: 45 kg/ha as basal and gypsum @ 200 kg/ha (100 kg at peg formation stage). For effective weed management, pre-emergence application of pendimethalin 0.75 kg/ ha followed by hand weeding at 45 DAS.

Winter Irrigated Cotton

Prepare the land with disc ploughing once + cultivator twice and form the ridges and furrows. For seed rate and spacing of variety (7.5 kg/ha with 75 x 30 cm) and for hybrid (1.0 kg/ha with 120 x 60 cm).The recommended nutrient management is 80: 40: 40 kg/ha for variety and 120: 60: 60 kg/ha for hybrid. For foliar application, spray 2 per cent $MgSO_4$ + 1 per cent urea twice during boll formation stage. Irrigation should be given at sowing, life irrigation on 5th DAS. Thereafter irrigate at 10 days interval (Red and laterite soils) or at 15 days interval (Black and Alluvial soils) till maturity. Application of pendimethalin or fluchloralin 1.0 kg/ha on 3 DAS + power weeding on 45 DAS followed by earthing up (or) Trifloxy sulfuron @10 g/ha on 15 DAS for effective control of broad leaved weeds and sedges. Intercultural operations should be done like topping at 15th node for varieties and 20th node for hybrids.

Winter Rainfed Cotton

Prepare the land with disc ploughing once + cultivator twice (for in-situ soil moisture conservation compartmental bunding/broad beds and furrows). For seed treatment, seed hardening with 2 per cent KH_2PO_4 or 2 per cent KCl is done with the seed rate of 15 kg/ha. Sowing should be done with seed drill. For nutrient management, apply NPK at 40: 20: 40 kg/ha. Need based foliar spray of 2 per cent $MgSO_4$ + 1 per cent urea twice during boll formation stage. For managing of weeds, application of pendimethalin 0.75 kg/ha on 3rd day after receipt of first rainfall followed by one hand weeding on 45 DAS or Trifloxy sulfuron @10 g/ha on 15 DAS for broad leaved weeds and sedges. Intercultural operation like Kaolin 3 per cent spray twice at peak flowering and boll formation stage. For moisture conservation, application of coirpith through tractor operated coirpith applicator.

Sugarcane

For preparation of land, form the ridges and furrows with a spacing of 80 cm between rows/paired row planting/pit method. For sett management, sett rate used are50,000 three budded setts/ha (or) 75,000 two buded setts/ha (or) 1, 87,500 single budded setts/ha and the setts are treated with Azophos 10 packets/ha of setts.The spacing to be adopted is paired row : 8 two budded setts per metre length of furrow, for mechanization 120 cm row space and double side planting and for pit method 32 single budded setts per pit. The recommended dose of fertilizer to be adopted is 300:100:200 NPK kg/ha 100 per cent P as basal. N and K in five equal splits at 30, 60, 90, 120 and 150 days after planting (DAP). For managing of weeds, application of Atrazine 2.0 kg/ha on 3rd DAP. For striga management, pre emergence application of Atrazine 2.0 kg/ha on 3 DAP, hand weeding on 45 DAP followed by earthing up on 60 DAP. Spray 2,4 – D Sodium salt 6 g or urea 200 g/t of water on 90 DAP and mulching with sugarcane trash @ 5 t/ha on 120 DAP. Irrigate once in 7 days (0-35 DAP) 10 days (36-100 DAP) 7 days (101 – 270 DAP) and 15 days up to maturity. Partial earthing up on 45th DAP followed by a complete earthing up on 90th and high level earthing up on 150th DAP. Detrashing should be done at 150th and 210th DAP with sugarcane detrasher (TNAU model) reduces the labour cost by 50 per cent. Propping should be done at 210th DAP.

Organic Farming

"An ecological production management system that promotes and enhances biodiversity, biological cycles and soil biological activity. It is based on minimal use of off-farm inputs and on management practices that restore, maintain and enhance ecological harmony."

WHO defines "organic agriculture as holistic food production management systems, which promotes and enhances agro-ecosystem health, including biodiversity, biological cycles and soil biological activity. It emphasizes the use of management practices in preference to the use of off-farm inputs, taking into account that regional conditions require locally adapted systems. This is accomplished by using, where possible, agronomic, biological and mechanical methods, as opposed to using synthetic materials, to fulfill any specific function within the system".

Principles of Organic Farming

- ☆ Organic production of crops and livestock and the management of farm resources so that they harmonize rather than conflict with natural systems.
- ☆ Use and development of appropriate technologies based upon an understanding of biological systems.
- ☆ Achieve and maintain soil fertility for optimum production by relying primarily on renewable sources.
- ☆ Use diversification to pursue optimum production.
- ☆ Aim for optimum nutritional value of staple food.
- ☆ Use decentralized structure for processing, distributing and marketing of products.
- ☆ Create a system which is aesthetically pleasing for those working in this system and those viewing it from out side.
- ☆ Maintain and preserve wildlife and their habitats.

Basic Concepts of Organic Farming

Organic farming system in India is not new and is being followed from ancient time. It is a method of farming system which primarily aimed at cultivating the land and raising crops in such a way, as to keep the soil alive and in good health by use of organic wastes (crop, animal and farm wastes, aquatic wastes) and other biological materials along with beneficial microbes (biofertilizers) to release nutrients to crops for increased sustainable production in an eco friendly pollution free environment.

Organic farming seeks to re-establish the balance of energy in nature without using chemical fertilizers and pesticides. This is mainly based upon traditional methods derived on sound ecological principles. Green revolution has turned into greed revolution and started posing several environmental problems. The agricultural scientists, environmentalists and policy makers are now proposing eco-friendly agriculture with slight modification of allowing judicious use of chemical fertilizers. The environment friendly agriculture does not allow any anthropogenic agricultural activities beyond the limit of threshold and it is mainly aimed to achieve optimum agricultural production without posing any severe problem to our environment. The traditional agriculture was based on organic farming system which manifested low crop and animal productivity but high environmental sustainability, because it was dependent on the use of organic inputs evolved naturally at the farm, animal power and maintaining the soil fertility and ecological balance, keeping the environment clean and providing stability to the food production without polluting our environment.

Integrated Farming System

Importance

Very often, almost all Indian farmers, in pursuit of supplementing their needs of food, fodder, fuel and finance, resort to adopt integrated farming systems, majority of them revolving around the crops and livestock components. Livelihood of small and marginal farmers, comprising 84 per cent of total farmers, depends mainly on crops or livestock, which is often affected by weather aberrations. An important consequence of this has been that their farming activities remain, by and large, subsistent in nature rather than commercial and many a times uneconomical. Under present scenario, in the absence of scientifically designed, economically profitable and socially acceptable appropriate integrated farming systems models, they are unable to harness the benefits of integration. Integration of different agriculturally related enterprises with crops provides ways to recycle the products and by-products of one component as input to another and reduce the cost of production and increase the total income of the farm.

Integrated Farming System - A Way to Sustainability

- ✰ In Integrated farming System, organic supplementation through effective utilization of residues/waste of linked components as manures. In addition vermicomposting constitutes an essential component of Integrated Farming System. Thereby IFS helps to maintain sustainable status of soil fertility in terms of physical, chemical and micro biological properties.
- ✰ Migration of rural poor and agricultural labors could be solved by integrating allied appropriate components in different eco zones as suited to varied resource situations. The rural and farm women falling under

small and marginal categories as well as Agricultural landless laborers will be benefited through the regular employment by integrated farming system.

- ✰ The integration of small ruminants and buffaloes, agro forestry and silvi-pastoral system along with cropping provide good scope for livelihood even under erratic rainfall years
- ✰ Optimal crop and livestock mixture consistent with the farm resources immediately available provides an opportunity to increase profitability and regular flow of income by virtue of intensification of crops and allied enterprises.
- ✰ Integrated farming system meets spread out demand for food, income and diverse requirements of food grains, vegetables, milk, egg, meat *etc.*, thereby improving the nutrition of small - scale farmers with limited resources.

Successful Integrated Farming Systems

In the last three decades Scientists at TNAU have conducted several research programs addressing various facets of integrated farming systems and have developed location specific viable models for different agro-climatic situations. On-farm integrated farming system models developed at TNAU, Coimbatore for wetland, irrigated land and dry land situations for effective utilization of farm resources and for sustained income generation are presented here.

a) Wetland ecosystem (1.00 ha)-Cropping (0.75 ha) + Fodder Crop (0.12 ha) + Fishery (1000 nos.) + Goat (20 female + 1 male) + Vermicompost (4t production Capacity)

Integration of crop with fish, poultry, pigeon and goat resulted in higher productivity than cropping alone under lowland. Crop + fish + goat integration recorded higher rice grain equivalent yield of 39610 kg/ha than other systems. Similarly, as an individual animal component, the goat unit (20 + 1) gave the highest productivity of 8818 kg. This could also provide 11.0 t of valuable manure apart from supplementing the feed requirement of 400 numbers of fish. The highest net return of ₹131118 and per day return of ₹511 ha^{-1} were obtained by integrating goat + fish + cropping applied with recycled fish pond silt enriched with goat droppings. The employment opportunity was also increased to 576 man days ha^{-1} $year^{-1}$ by integrating fish + goat in the cropping as against cropping alone (369 man days ha^{-1} $year^{-1}$). Combining cropping with other allied enterprises would increase labour requirement and thus provide scope to employ family labour round the year.

b) Irrigated upland ecosystem (1.00 ha) - Cropping (0.70 ha) + milch cows (3 + 2) + Goat (10 female+ 1 male) + Vegetables (0.06 ha) + Fodder Crop (0.20 ha) + Vermicompost (6t production Capacity)

The highest productivity of 23328 kg ha^{-1} as maize grain equivalent yield, net return of ₹ 64125 ha^{-1}, employment generation of 235 man days $acre^{-1}$ were obtained by integrating dairy, goat, guinea fowl, vermicompost and cropping applied with recycled manure to the first crop in the system and were higher than the traditional farming system. Additional benefit of 47.9, 40.1and 44.7 per cent over traditional farming practices in terms of productivity, profitability and employment were obtained. In addition, 7.83 tonnes per ha of crop and animal residues could be recycled through effective recycling.

c) Dry land eco system (1.00 ha)-Silivipasture (*Cenchrus setigerus* + *Stylosanthes hamata* and fodder sorghum + Pillipesara) + sheep (5+1) + buffalo (2 No.'s)+ vermicompost (5 t production Capacity)+ farm pond

Cenchrus setigerus + *Stylosanthes hamata* and fodder sorghum + Pillipesara system with sheep (5+1) and buffalo (2 No.'s) was promising, which generated the highest system productivity of 67660 kg of *Cenchrus* equivalent yield with net return of 32485 ha^{-1} $year^{-1}$ and benefit cost ratio of 2.58 with an employment opportunity of 169 man days ha^{-1} $year^{-1}$.*Cenchrus setigerus* + *Stylosanthes hamata* and fodder sorghum + Pillipesara with the application of recommended dose of NPK for individual crops along with FYM would be the best for obtaining higher green and dry fodder yields, crude protein yield, nutrient uptake, net return and benefit cost ratio. Integrated farming system's cafeteria approach has many options matching with farmers' preference, resource availability and affordability.

References

1. Chinnusamy, C., Prabhakaran, N.K., Janaki, P. and Govindarajan, K. 2008. Weed science research in Tamil Nadu. 25 years compendium, Department of Agronomy, TNAU, Coimbatore-3.
2. Jayanthi, C., P. Devasenapathy and C. Vennila. 2008. Farming system- Principles and Practices (ISBN No. 81-89304-49-6), Satish Serial Publishing House, New Delhi.
3. Jayanthi, C., S. Mythili, M. Balusamy, N. Sakthivel and N. Sankaran. 2003. Integrated nutrient management through residue recycling in lowland integrated farming systems. Madras Agri. J. 90(1-3): 103-107.
4. Jayanthi, C., M. Balusamy, C. Chinnusamy and S. Mythili. 2003. Integrated nutrient supply system of linked components in lowland integrated farming system. Indian Journal of Agronomy. 48(4): 241-246.
5. Jayanthi,C., C. Vennila, K. Nalini and B.Chandrasekaran. 2009.Sustainable integrated management of crop with allied enterprises: Ensuring livelihood security of small and marginal farmers. Asia Pacific Tech Monitor. Jan – Feb issue. Pp. 21-27.
6. Jayanthi,C., C. Vennila and K. Nalini. 2008. Farmer's participatory research on integrated farming system. In: proceedings of the 14th Australian Conference. Adelaide South Australia.

7. Murugappan.V. 2000. Integrated Nutrient Management- The concept and Overview. Theme paper on Integrated Nutrient Management. Published by Tamil Nadu Agricultural University and Tamil Nadu Department of Agriculture.

8. Natrajan. S., R.Venkitaswamy, R.Jagadeeswaran and C.Sudhalakshmi.2007. Crop Management Technologies. Sri Sakthi Promotional Litho Process, Coimbatore.

Chapter 5

Integrated Nutrient Management: One Stop Solution for Sustainable Soil Health, Crop Productivity and Removal of Soil Fatigue

V.V. Krishnamurthi, T. Chitdeshwari and R. Santhi

Department of Soil Science and Agricultural Chemistry
Tamil Nadu Agricultural University,
Coimbatore – 641 003, Tamil Nadu

Globally food demand is expected to be doubled by 2050 and projections showed that demand for food grains in India would increase from 192 to 345 million tonnes in 2030. Therefore, in the next 20 years, production of food grains needs to be increased at the rate of 5.5 million tonnes annually (ICAR, 2011). Limited availability of additional land for crop production coupled with declining yield in major food crops, heightened many concerns about agriculture's ability to feed the world population which is expected to exceed 7.50 billion by the year 2020 (Subba Rao and Sammi Reddy, 2005; FAO, 2012). Decreasing soil fertility has also raised concerns about the sustainability of agricultural production at current levels.

The common barriers and root causes for the decline in agricultural productivity, soil fertility and the ensuing threat to food security in Asian countries are:

- Soil mining of nutrient reserves and Poor rationalization of chemical fertilizers
- Inadequate understanding on balanced fertilization based on soil test crop response approach
- High population and rapid urbanization and declining man to land ratio from 0.50 ha in 1951 to 0.14 ha at present and is expected to decline further to 0.08 ha by the end of 2020
- Declining total factor productivity of all major crops due to wider nutrient demand and supply
- Emerging secondary and micronutrients deficiencies
- Gradual decrease in soil organic matter and an increase in soil degradation (erosion, acidification, salinization, alkalization pollution, compaction *etc.*)
- Uncontrolled increase in the price of chemical fertilizers
- Soil erosion resulting from decreasing fallow periods, subsistence farming and increasing human demand for land and food.

Concept of Integrated Nutrient Management

Integrated Nutrient Management (INM) is a strategy to maintain soil fertility for sustaining increased crop productivity through optimizing all possible sources of organic and inorganic plant nutrients required for crop growth and quality in an integrated manner, appropriate to each cropping system and farming situation in its ecological, social and economic possibilities (Roy and Ange, 1991). Therefore, it is a holistic approach integrates the use of all natural and man-made sources of plant nutrients, so that crop productivity increases in an efficient and environmentally benign manner, without sacrificing soil productivity for future generations (Gruhn *et al.*, 2000, Chaudhari *et al.*, 2015).

Determinants of INM

- Nutrient requirement of cropping system as a whole
- Soil fertility status and special management needs to overcome soil problems
- Local availability of nutrients resources (organic, inorganic and biological sources)
- Economic conditions of farmers and profitability of proposed INM option
- Social acceptability
- Ecological considerations
- Impact on the environment

Components of INM

Conceptually, balanced fertilization would essentially mean rational use of fertilizers and organic manures for the supply of nutrients for crop production in a manner that would ensure (i) efficiency of fertilizer use, (ii) harnessing best possible positive and synergistic interactions amongst the various other factors of crop production (seed, water, agrochemicals, *etc.*), (iii) least adverse effect on environment by minimizing nutrient losses (iv) maintaining soil productivity and (v) sustaining high yield commensurate with the biological potential of the crop under unique soil, climate and agro-ecological set up (Zhang *et al.*, 2012). General or blanket fertilizer recommendations are not based on soil fertility and may lead either under or over usage of fertilizers.

Organic Manures

Organic manures such as FYM, crop residues, composts, vermicompost, biogas slurry, bio-compost, pressmud, oil cakes, biofertilizers, etc has the potential of supplying not only the plant nutrient but also improves soil organic matter and biological activity. In India, around 350 Mt of organic wastes was generated and the total nutrient (NPK) potential of various organic resources was estimated to be 14.85 Mt in 2000, which would become 32.41 Mt by 2025 (Katyal, 2015). About 8.9 Mt of pressmud cake and 12.1 Mt of poultry manure are available annually in India. The present FYM availability in the country is estimated to be around 1.70 t ha^{-1} of net cultivated area and the nutrient content of various organic/animal manures and crop residues available in India is given in Table 5.1.

Table 5.1: Nutrient Potentials of Organic Sources

	Nutrient Content (per cent)		
	Nitrogen (N)	*Phosphorus (P_2O_5)*	*Potassium (K_2O)*
	Manure		
Farmyard manure	0.40-1.50	0.30-0.90	0.30-1.90
Coir pith	1.20	1.20	1.20
Poultry manure, fresh	1.0-1.8	1.40-1.80	0.80-0.90
Press mud	1-1.5	4-5	2-7
Raw bone meal	3-4	20-25	-
Blood meal	10-12	1.20	1.00
Fish meal	4-10	3.9	0.3-1.5
Animal refuse	0.3-0.4	0.1-0.2	0.1-0.3
Sewage sludge, dry	2.0-3.5	1.0-5.0	0.2-0.5
	Oil cakes		
Neem cake	5.20	1.00	1.40
Coconut cake	3.00	1.90	1.80
Groundnut cake	7.30	1.50	1.3
Sesame cake	6.20	2.00	1.20
	Crop Residues		
Groundnut husks	1.6-1.8	0.3-0.5	1.1-1.7
Banana	0.61	0.12	1.00
Cotton	0.44	0.10	0.66

	Nutrient Content (per cent)		
	Nitrogen (N)	Phosphorus (P_2O_5)	Potassium (K_2O)
Paddy	0.36	0.08	0.71
Wheat	0.53	0.10	1.10
Sugarcane trash	0.35	0.10	0.60
Tobacco dust	1.10	0.31	0.93

Source: Jat *et al.*, 2015.

Crop residues received from major crops is about 234 Mt/year which is 34 per cent of gross residue generated in India however only 5 per cent of the crop residues are presently recycled in agriculture and the remaining residues are consumed as cattle feed or burnt. Application of organic manures generally increases crop yields compared to fertilizers addition alone and the increase in yield potential is due to many components in the organic manures and their effects on improved soil structure, water regime and trace element supply. Organic manures also have considerable potential in increasing the use efficiency of chemical fertilizers by crops. Slow nutrient release from manures and composts provides stable supply of NH4 and thus support maximal yields (Kaur *et al.*, 2009).

a. Biofertilizers

Biofertilizers are N-fixing, mineral solubilization, cellulolytic microorganisms facilitate economizing fertilizer nutrient use through solubilizing less mobile nutrients and recycling nutrients from crop residues. Integration of such systems makes the production system more stable and sustainable (Meena *et al.*, 2013). The global market for biofertilizers in terms of revenue was about 5 billion USD in 2011 and forecasted to double by 2017 (Jat *et al.*, 2015). However the production in India is around 50,000 tonnes per year though the potential requirement is much higher. Larger benefits associated with the bio fertilizers like required in small quantities, increasing nutrient use, cheap and reducing the dose of chemical fertilizers attracted the use of biofertilizers for crop production by the farming communities. Commonly used bio-fertilizers in India are mentioned below along with some salient features (Table 5.2).

INM on Soil Health

Long term experiments in the country and elsewhere clearly demonstrated a significant increase in soil organic carbon content due to the addition of NPK + FYM followed by NPK alone in various crops and cropping systems (Table 3).

Maintaining the soil organic carbon (SOC) stock is the most potent weapon in fighting against soil degradation and ensuring sustainability in agriculture. Results of long- term fertiliser experiments, conducted for more than 40 years, proved beyond doubt that inclusion of organic manures in the crop production is essential for sustaining the productivity of crops besides improving the SOC stock (Arulmozhiselvan et, al, 2011;Srinivasarao *et al.*, 2012,).

Table 5.2: Benefits of Commonly Used Bio-fertilizers

Bio-fertilizers			*Benefits*	*Crops*
N fixers	Rhizobium	Ability to fix atmospheric nitrogen in symbiotic association with plants by forming nodules in roots	Fixes 20-50 kg N per ha 10-35 per cent yield increase	Pulses and Legumes
	Azatobacter	Free living and non-symbiotic nitrogen fixing organism	Fixes N produces growth promoting substances Suppress many root pathogens	Cereals, millets, vegetables, cotton and sugarcane
	Azospirillum	Nitrogen-fixing micro organism beneficial for non-leguminous plants.	Fixes 20-25 kg N per ha Produce growth promoting substances	Cereals, vegetables, cotton, Sugarcane
P solubilizers	Phosphate solubilizing bacteria (PSB)	Solubilizes insoluble P compounds in soil *Pseudomonas, Bacillus, Rhizobium, Burkholderia, Achromobacter, Agrobacterium, Microccocus, Aereobacter Flavobacterium*	Solubilise insoluble P 10-20 per cent yield increase Produce growth promoting substances	All crops
	VAM	Fungi roots has a symbiotic association with host plants	Mobilizes P Enhances the uptake of P, Zn, S and Water Protects against pathogens	All crops
Zn solubilizers	*Bacillus subtilis, Thiobacillus thioxidans* and *Saccharomyces* sp	Solubilise native insoluble Zn compounds such as zinc oxide, zinc carbonate, zinc sulphide *etc.* and made available to the crops	Increases yield and solubilise insoluble Zn	Rice

Source: Natarajan *et al.* (2000), Misra *et al.* (2013); Mohd Mazid and Taqi Ahmed Khan (2014).

Table 5.3: Organic Carbon Content under Long-term Fertilizer Use

Soil Type	*Cropping Sequence*	*Soil Organic Carbon (per cent)*			
		Initial	*Unfertilised*	*NPK*	*NPK+FYM*
Inceptisol	Rice - Rice	0.27	0.41	0.59	0.76
Inceptisol	Rice-Wheat	0.37	0.19	0.40	0.50
Inceptisol	Maize-Wheat	0.44	0.32	0.47	0.56
Mollisol	Rice-Wheat	1.48	0.50	0.95	1.51
Chromustert	Soybean-Wheat-Maize	0.57	0.53	0.62	0.97
Alfisol	Maize-Wheat	0.79	0.62	0.83	1.20

Source: Singh *et al.*, 2012.

Subba Rao and Sammi Reddy (2005) and Santhi *et al.* (2012) compiled and critically examined importance and relevance of IPNS and the opportunities for managing IPNS for various crops in different agro-eco regions of India (Table 5.4).

Table 5.4: INM Strategies for Major Crops and Cropping Systems

Cropping System	*INM Strategy*
Rice-rice	Use of organic sources like FYM, compost, green manure, azolla etc, meet 25-50 per cent of N needs in Kharif rice and can help curtailing NPK fertilizer by 25-50 per cent
	Application of 75 per cent NPK as fertilizers + 25 per cent NPK through green manure or FYM to rice is optimal for increasing yield and sustaining soil health.
	A successful inoculation of blue green algae @ 10kg/ha provides about 20-30 kg N/ha
Rice- Potato-Groundnut	Use 75 per cent NPK with 10 t FYM/ha to rice and potato.
Sugarcane based cropping system	Combined use of 10 t FYM/ha and recommended NPK increases the cane productivity by 8-12 t/ha over chemical fertilizer alone
Pulses	Integrated use of FYM and 50 per cent recommended NPK fertilizer plus rhizobium inoculation helps in saving 25 per cent chemical fertilizer
Sorghum	Substitution of 60 kg N through FYM or green loppings of Leucaena leucocephala increases the yields and FUE.
Cotton	50 per cent of recommended NPK can be replaced by 5 t FYM/ha
Oilseeds (Mustard, Sunflower etc.)	Substitution of 25-50 per cent of chemical fertilizer through 10 t FYM/ha increases the yield and FUE

Soil physical properties are closely related with soil organic matter, thus, any soil management practice which enhances soil organic matter has direct bearing on soil physical properties and microbial biomass. Significant improvements in soil physical properties were observed by many researchers under INM. Addition of NPK fertilizers along with organic manure and biofertilizers increased soil moisture-retention capacity and infiltration rate while reducing the bulk density (Saha *et al.*, 2010). Incorporation of organic has a significant effect on soil aggregation (Saikia *et al.*, 2015), soil structure, soil moisture-retention capacity (Abdollahi *et al.*, 2014) and infiltration rate (Sharma *et al.*, 2015). Several researchers reported an increase in soil physical properties due to the combined addition of organic manures and inorganic fertilizers (Table 5.5).

Table 5.5: Effect of Integrated Nutrient Management Practices on Soil Physical Properties

Crops	*Response to INM Practices*	*References*
Soybean, Wheat, maize, Wheat-Soybean	Decreased the bulk density and particle density, increased soil organic carbon,	Azis *et al.*, 2015, Kushro *et al.*, 2013, Kannan *et al.*, 2013, Battacharya *et al.*, 2007
Maize, Cotton-Wheat, French bean, Pea-Wheat, Rice-Wheat	Improved total porosity, hydraulic conductivity, soil moisture content	Nwite *et al.*, 2014, Hassan *et al.*, 2013, Datt *et al.*, 2013, Liu *et al.*, 2013, Rasool *et al.*, 2007
Tomato	Reduced soil temperature, BD significantly reduced	Rasool *et al.*, 2007

Most of the research results clearly demonstrated that INM enhances the yield potential of crops over and above the achievable yield with recommended fertilizers and results in better synchrony of crop nutrient needs (Table 5.6).

Judicious application of mineral fertilizers and organic manure along with biofertilizers and micronutrients gave highest available NPK in soil as compared to other treatment combinations (Stalin *et al.*, 2010, Meena *et al.*, 2013). Incorporation of FYM, green manures and BGA coupled with inorganic fertilisers increased the organic carbon, mineralisable nutrients and their availability (Kumar *et al.*, 2012)

Inclusion of biofertilisers in the nutrient schedule not only improves the crop yield and soil nutrient availability but also increased the microbial population in soil. The buildup of micro flora in soil due to biofertilizer addition as a complimentary effect was reported by many researchers (Table 5.7).

Santhi *et al.* (2010a) documented soil test based fertilizer prescriptions developed under IPNS on various soil types for various crops in different agroclimatic zones of Tamil Nadu. Accomplishing all these recommendations, computer software 'DSSIFER' (Decision Support System for Integrated Fertilizer Recommendation) was developed to generate crop, site and situation specific balanced fertilizer prescriptions for Tamil Nadu and has been revised and updated as DSSIFER 2010 (Santhi *et al.*, 2010b). Further, to identify the nutrient deficiencies in crops, a computer software "Visual Diagnostic Kit" has been developed by TNAU (Subramanian *et al.*, 2008). These softwares can be used as smart tools for promoting IPNS and maintaining soil health and sustainable productivity.

Advantages of INM

- ✰ Enhances the availability of applied as well as native soil nutrients
- ✰ Synchronize the crop nutrient demand and nutrient supply from native and applied sources

Table 5.6: Effect of INM on Soil Fertility and Crop Productivity

Crops	*Response to INM*	*References*
Wheat-Maize, Pea-French bean Wheat, Pea-Wheat	Significantly increased soil organic carbon, total N, enzymatic activities, microbial colonization	Liang *et al.*, 2014 Mahanta *et al.*, 2013, Jat *et al.*, 2013 Hemalatha *et al.*, 2010
Rice-Rice, Rice-Barley	Significantly increased in NUE and soil organic carbon, enzymatic activity	Liang *et al.*, 2003, Xu *et al.*, 2008
Cereal-Legume	Green manure with mineral fertilizer ensure higher crop productivity, soil fertility	Rahman *et al.*, 2013
Maize	RDF + Vermicompost enhances NPK availability and microbial activity	Kumar *et al.*, 2012, Kannan *et al.*, 2013
Chili	Highest available NPK and micronutrients, higher yield	Naidu *et al.*, 2009
Rice, Green gram	Enhances available NPK in post harvest soil	Cho *et al.*, 2003 Lakshmi *et al.*, 2014
Cotton	INM significantly increased NPK uptake and sustain soil fertility	Marimuthu *et al.*, 2014
Finger Millet, Maize, Cowpea	Combined application of 100 per cent NPK + 10 t FYM ha^{-1} significantly increased the yield by 27 to 39 per cent, nutrient availability, their use efficiency, and organic matter content of the soils	Dhakshinamoorthy *et al.*, 2005,
Sunflower, Maize, Sorghum, Cotton, Ragi, Tenai, varagu, panivaragu, Wheat, Cumbu,	In the Permanent Manurial experiments conducted for more than 100 years revealed that application of 100 per cent NPK+ 12.5 t FYM ha^{-1} increased the yield of crops under irrigated and rainfed condition besides improving the soil health	Arulmozhiselvan *et al.*, 2011
Rice, Sunflower	RDF+2.5 kg Zn as enriched FYM increases the yield by 6 - 35.7 per cent	Stalin *et al.*, 2010
Maize, Groundnut	RDF+25 kg $ZnSO_4$ as enriched FYM increases the yield by 15 per cent	
Rice	RDF + 37.5 kg $ZnSO_4$ + 5 t daincha + gypsum at 50 per cent GR ha-1 increased the yield in sodic soils	
Sugarcane, Sorghum, Maize	RDF+ 100 kg $FeSO_4$ + 12.5 t FYM increases the yield and Fe use efficiency	

Table 5.7: Effect of INM and Biofertilizer on Crop Performance and Soil Health

Crops	*Response*	*Reference*
Maize	Bioinoculants with 50 per cent RDF showed a significant yield increase	Sumagayasay, 2014
Lentil	FYM and biofertilizer improve soil health	Moraditochaee *et al.*, 2014
Soybean Sunflower Mungbean	RDF with biofertilizer resulted in improved productivity, soil fertility and nutrient balance	Gharpinde *et al.*, 2014 Akbari, *et al.*, 2011 Rana *et al.*, 2011
Groundnut, Soybean, Greengram, Cowpea, Blackgram and Redgram	Rhizobium inoculation showed11-65 per cent yield increase in crops Residual effect on succeeding crops also reported	Narayanan *et al.*, 1996 Natarajan *et al.*, 2000 Kannaiyan *et al.*, 2000
Millets, Cotton, Gingelly	Azospirillum inoculation increased the yield of crops by 16-30 per cent	Natarajan *et al.*, 2000
Rice	Incorporation of 10 t fresh Azolla saved 30 kg N ha-1	Natarajan *et al.*, 2000 Kannaiyan *et al.*, 2000

- Provides balanced nutrition to crops and minimizes the antagonistic effects resulting from hidden deficiencies and nutrient imbalances
- Improves and sustains the physical, chemical and biological functioning of soil
- Minimizes the deterioration of soil, water and ecosystem by promoting carbon sequestration, nutrient toxicities, *etc.*

References

Abdollahi, L., Schjonning, P., Elmholt, S., Munkholm, L.J. (2014). The effects of organic matter application and intensive tillage and traffic on soil structure formation and stability. Soil Till. Res., 136: 28-37.

Akbari, P., Ghalavand, A., Modarres, S., A.M. and Alikhani, M.A. (2011). The effect of biofertilizers, nitrogen fertilizer and farmyard manure on grain yield and seed quality of sunflower (Helianthus annus L.). J. Agril. Tech., 7 (1): 173-184.

Aziz, M.A, Mushtaq, T., Ali, T., Islam, T. and Rai A.P. (2015). Effect of integrated nutrient management on soil physical properties using soybean (Glycine max (L.) Merill) as indicator crop under temperate conditions. J. Envir. Sci., Comp. Sci. and Eng.Techn., 4: 56-64.

Bhattacharyya, R., Chandra, S., Singh, R.D., Kundu, S., Srivastva, A.K. and Gupta, H.S. (2007). Longterm farmyard manure application effects on properties of a silty clay loam soil under irrigated wheat–soybean rotation. Soil and Tillage Res., 94: 386-396.

Chaudhari,S.K.Adlul Islam,P.P.Biswas and A.K.Sikka (2015). Integrated soil, water and nutrient management for sustainable agriculture in India. Indian J.Fertilisers, 11(10): 51-63.

Cho, B.J.Y., Son, J.G., Song, C.H., Hwang, S.A., Lee, Y.M., Jeong, S.Y., Chung, B.Y. (2008). Integrated nutrient management for environmental-friendly rice production in salt-affected rice paddy fields of Saemangeum reclaimed land of South Korea. Paddy Water Environ., 6: 263-273.

Datt N., Dubey, Y. P.and Chaudhary, R. (2013).Studies on impact of organic, inorganic and integrated use of nutrients on symbiotic parameters, yield, quality of french-bean (Phaseolus vulgaris l.) vis-à-vis soil properties of an acid Alfisol. African J. Agric.Res., 8 (22): 2645-2654.

Dhakshinamoorthy, M.V.Singh, P.Malarvizhi, D.Selvi and A.Bhaskaran (2005). Soil quality, crop productivity and sustainability as influenced by long term fertilizer application and continuous cropping of finger millet – maize – cowpea sequence in swell shrink soils, Research bulletin No.3, published by ICAR-AICRP and TNAU, Pages, 1-122.

Food and Agriculture Organization (FAO), (2012). World Agriculture Towards 2030/2050: The 2012 revision ESA E Working Paper No. 12-03, 2012.

Gharpinde, B., Gabhane, V.V., Nagdeve, M.B., Sonune, B.A. and. Ganvir, M.M.(2014). Effect of integrated nutrient management on soil fertility, nutrient balance, productivity and economics of soybean in an Inceptisol of semi arid region of Maharashtra. Karnataka J. Agric. Sci., 27 (3): 303-307.

ICAR, (2011). Vision 2030. Indian Council of Agricultural Research, New Delhi.

Jat, M.L. Majumdar, K. McDonald, A. Sikka, A.K. and Paroda, R.S. (2015). Book of extended summaries. National Dialogue on Efficient Nutrient Management for Improving Soil Health, September 28-29, 2015, New Delhi, India, TAAS, ICAR, CIMMYT, IPNI, CSISA, FAI, p. 56.

Katyal, J.C. (2015). Soils through the lens of prakriti, sanskriti, Niyat and Niti. Indian Journal of Fertilisers 11(10), 17-51.

Kusro, P.S., Singh, D.P., Paikra, M.S. and Kumar, D.(2014). Effect of organic and inorganic additions on physico-chemical properties in Vertisol. Amer. Int. J. Res. Formal, Applied and Nat. Sci., 5 (1): 51-53.

Lakshmi, C.S. R., Rao, P.C., Sreelatha, T., Padmaja, G. Madhavi, M., Rao, P.V. and Sireesha, A. (2014). Residual effects of INM on humus fractions, micronutrient content and their uptake by rabi greengram under rice-pulse cropping system. Res. Crops, 15 (1): 96-104.

Liang, Q., Chen, H., Gong, Y., Yang, H., Fan, M.and Kuzyakov, Y. (2014). Effects of 15 years of manure and mineral fertilizers on enzyme activities in particle-size fractions in a North China Plain soil. Europ. J. Soil Bio., 60: 112-119.

Mahanta, D., Bhattacharyya, R., Gopinathc, K.A., Tuti, M.D., Jeevanandan, K., Chandrashekara, C., Arunkumar, R., Mina, B.L., Pandey, B.M., Mishra, P.K., Bisht, J.K., Srivastva, A.K. and Bhatt, J.C.(2013). Influence of farmyard manure application and mineral fertilization on yield sustainability, carbon sequestration potential and soil property of Garden pea–French bean cropping system in the Indian Himalayas. Scient. Horti., 164: 414-427.

Marimuthu, S., Surendran, U., Subbian, P. (2014). Productivity, nutrient uptake and postharvest soil fertility as influenced by cotton-based cropping system with integrated nutrient management practices in semi-arid tropics. Arch. Agron. Soil Sci., 60: 87-101.

Moraditochaee, M., Azarpour, E. and Bozorgi, H.R. (2014). Study effects of bio-fertilizers, nitrogen fertilizer and farmyard manure on yield and physiochemical properties of soil in lentil farming. Int. J. Biosci., 4 (4): 41-48.

Mosa, W.F.A. El-Gleel, Paszt, L.S., Frcl, M. and TrzciDski P. (2015) The Role of Biofertilization in Improving Apple Productivity A Review. Adv. Microbio., ; 5: 21-27.

Mohd Mazid and Taqi Ahmed Khan (2014). Future of Bio-fertilizers in Indian Agriculture: An Overview, International Journal of Agricultural and Food Research, 3, 10-23.

Rahman, M.H., Islam, M.R., Jahiruddin, M., Rafii, M.Y., Hanafi, M.M. and Malek, M.A. (2013). Integrated nutrient management in maize-legume-rice cropping pattern and its impact on soil fertility. J. Food Agric. Environ. 11: 648-652.

Saikia, P., Bhattacharya, S.S. and Baruah, K.K. (2015).Organic substitution in fertilizer schedule: Impacts on soil health, photosynthetic efficiency, yield and assimilation in wheat grown in alluvial soil. Agri. Eco. and Environ. 203: 102–109.

Sharma, V.K., Pandey, R.N. and Sharma, B.M. (2015). Studies on long term impact of STCR based integrated fertilizer use on pearl millet(Pennisetum glaucum) -wheat (Triticum aestivum) cropping system in semi arid condition of India. J. Environ. Bio., 36: 241-247.

Sumagaysay, C.L. (2014). Bio N fertilization on Corn. Int. J. Biotech., 3 (6): 85-90.

Chapter 6

Sustainable Agriculture through Effective Integrated Pest Management

T. Manoharan and S. Jeyarani

Department of Agricultural Entomology
Tamil Nadu Agricultural University,
Coimbatore – 641 003, Tamil Nadu

The steady increase in human population and shrinking land area are exerting a severe pressure on Indian Agriculture. Intensive agriculture involving fertilizer responsive high yielding varieties and monocropping have resulted in the outbreak of several crop pests. Annually, we have been losing nearly 50 per cent of our production due to these limiting factors. A wide range of recommended chemical insecticides are being used by the farmers for controlling the pests. The chemicals are highly effective, rapid in action, adaptable to most situations and relatively economical. Despite these advantages, the use of chemical pesticides had been ecologically unsafe and harmful to natural enemies. The use of persistent insecticides acquires special concern on vegetables with little time gaps between treatment and consumption. The increasing concern for environmental safety and global demand for pesticide free food necessitated the search for eco friendly methods of pest management and this lead to the origin of Integrated Pest Management (IPM) concept. IPM refers to an ecological approach in which all available, necessary, efficient and economic techniques are consolidated in a unified programme, so that pest populations can be managed in such a manner that economic damage is avoided and adverse side effects are minimized. The important components of IPM for the successful suppression of pests are discussed here.

I. Biological Control

a. Entomophagous Insects

Despite the slow action and specificity, biocontrol agent like pathogens, parasitoids, predators and antagonistic organisms have become invaluable components in agricultural IPM system in view of high level of specificity, safety and sustainability. The high level of human safety, stability of control and renewable nature, make them very attractive candidates for pest management. Some of the potential biocontrol agents for the management of pests of different crops are furnished in Table 6.1.

Table 6.1: Potential Biocontrol Agents for the Management of different Crop Pests

Crop	*Pest*	*Natural Enemy*	*Release Rate*
		Parasitoids	
Rice	Stem borer	*Trichogramma japonicum*	40,000 ac^{-1} thrice at weekly intervals
	Leaf folder	*Trichogramma chilonis*	40,000 ac^{-1} thrice at weekly intervals
Sugarcane	Early shoot borer	*Sturmiopsis inferens*	Release 50 gravid females of ac^{-1} on 30 and 45 DAP
	Internode borer	*Trichogramma* sp.	20,000 ac^{-1} thrice at fortnightly intervals
	Top borer	*Isotima javensis*	50 females ac^{-1}
	Pyrilla perpusilla	*Epiricania melanoleuca*	3200 to 4000 cocoon ac^{-1} (or) 3.2 to 4 lakhs egg ac^{-1}
Cotton	Pink boll worm	*Trichogrammatoidea bactrae*	40,000 ac^{-1} thrice at weekly intervals coinciding with pest incidence
	H. armigera	*T. chilonis*	2.5cc ac^{-1} thrice at fortnightly intervals from 45 DAS
		Chelonus blackburnii	40,000 ac^{-1}
		Predators	
Rice	Hoppers	*Cyrtorhinus lividipennis*	Natural predation
		Coccinella arcuata	
		Microvelia atrolineata	
		Geocoris bicolor	
Cotton	*H. armigera*	*Chrysoperla carnea*	40,000 ac^{-1} at 6th, 13th and 14th week after sowing
Brinjal	*Coccidohystrix insolitus*	*Cryptolaemus montrouzieri*	600 beetles ac^{-1}
Grapes and Guava	Mealybugs	*Cryptolaemus montrouzieri*	600 beetles ac^{-1}

b. Entomopathogens

Among the different microbial agents developed and tested, bacteria, viruses, and fungi are considered promising agents to control insect pests. In India, few entomopathogens that are being commercially used are furnished in Table 6.2.

Table 6.2: Potential Entomopathogens for the Management of different Crop Pests

Pathogen	*Crop and Pest*	*Recommended Dose*
	Virus	
Nuclear polyhedrosis virus (NPV)	Cotton - *H. armigera*	Apply NPV at 1.2 x 10^{12} POB ac^{-1} in evening hours at 7th and 12th week after sowing
	Cotton and Groundnut - *S. litura*	Apply NPV at 0.6 x 10^{12} POB ac^{-1} in evening hours
	Pulses - *H. armigera*	Apply NPV at 0.6 x 10^{12} POB ac^{-1} in evening hours
	Pulses - *S. litura*	Apply NPV at 0.6 x 10^{12} POB ac^{-1} in evening hours
Granulosis virus	Sugarcane early shoot borer, *Chilo infuscatellus*	Apply granulosis virus at 0.6 x 10^{12} POB ac^{-1} twice on 35 and 50 days after planting (DAP)
	Bacteria	
Bacillus thuringiensis	*S. litura* and *H. armigera*	0.4 kg ac^{-1}
	Diamond back moth, *Plutella xylostella*	0.4 kg ac^{-1} at primordial stage (ETL 2 larvae/plant)
	Rice stem borer, *Scirpophaga incertulas* and rice leaf folder, *Cnaphalocrocis medinalis*	0.4 kg ac^{-1}
	Fungus	
Beauveria bassiana	Cotton - *H.armigera, S.litura, Chilo partellus, Scirpophaga incertulus* etc.	*Beauveria bassiana* 1.15 per cent WP 160 g ac^{-1}
	Mango - hopper	*Beauveria bassiana* @ 10^{8} cfu ml^{-1} on tree trunk once during off season and twice at 7 days interval during flowering season
	Pulses - Root rot-stem fly complex	Seed treatment with *Beauveria bassiana* + *Pseudomonas fluorescens* @ 5 g each/kg of seed
	Cotton whitefly	*Verticillium lecanii* 1.15 per cent WP 1000 g ac^{-1}

Source: CPG Agri, 2012 and Horti, 2013.

II. Cultural

It aims at the manipulation of cultural practices to the disadvantage of pests.

a. Seeds and Sowing Dates

Pest free seed and planting materials should be used. Changes in sowing and planting dates can help in the avoidance of egg laying period of certain pests, the establishment of tolerant plants before pest occurrence, the maturation of crop before abundance of pests, the synchronization of pests with natural enemies.

Early sowing, increased seed rate (5 kg/ac) and use of fish meal trap @ 5 ac^{-1} till the crop is 30 days old, minimized the sorghum shoot fly incidence.

b. Habitat Diversification in Pest Management

Habitat diversification makes the agricultural environment unfavourable for the growth, multiplication and establishment of insect pest populations and also encourages natural enemies' population. Habitat diversification is normally achieved through practices such as crop rotation, polyculture, crop density, trap crops, tillage, water management, fertilizer management, organic mulches, plastic and reflective mulches, cover crops, planting times, short-season crops, sanitation, fallow, physical barriers and combinations.

i. Trap Cropping

Trap crops are the plants that are grown to attract insect or other organisms to protect target crops from pest attack. Pests can be strongly attracted by certain plants. When these crops are sown in the field or alongside it, insects will gather on them and can thus be easily controlled. For example, the cotton bollworm, *Helicoverpa armigera* prefers maize than cotton for egg laying. Hence, few rows of maize can be sown in cotton field to attract the moth for egg laying and then the eggs on these plants can then be killed by insecticides.

Growing of marigold to attract pests like American bollworm, barrier crops like maize/jowar to prevent migration of sucking pests like aphids and guard crops like castor to attract *Spodoptera litura* in cotton fields is also recommended.

ii. Inter-cropping

Interplanting maize in cotton fields increase the population of spiders, coccinellids and chrysopids and also acts as a trap crop for *H. armigera*. Intercropping pulses in cotton reduce the leaf hopper population on cotton and Lablab bean in sorghum reduce sorghum stemborer incidence.

Intercropping of groundnut with pearl millet reduce the incidence of thrips, jassid and leaf miner. Growing pearl millet as intercrop in groundnut increases the parasitic activity of *Goniozus* sp. Intercropping of tomato with cabbage (Cabbage planted 30 days later than tomato) reduces egg laying by diamond backmoth.

iii. Mixed Cropping

Mixed cropping helps in pest control many ways. Chemical signals from the mixed crop decreases pest incidence. It also influences the availability of light which can affect insect behaviour. Growing of lab-lab, cowpea or other short statured pulses as a mixed crop with sorghum minimizes infestation by stem borer.

Some of the cultural practices along with the pests that could be managed are presented in Table 6.3.

III. Host Plant Resistance (HPR)

Varieties/Hybrids with tolerance or resistance to insect pests can be effectively used to reduce the pest incidence.

Table 6.3: Crop Pests Checked by the Cultural Practices

Sl.No.	Cropping Techniques	Pest Checked
1.	Ploughing	Red hairy caterpillar
2.	Puddling	Rice mealy bug
3.	Trimming and plastering	Rice grass hopper
4.	Pest free seed material	Potato tuber moth
5.	High seed rate	Sorghum shoot fly
6.	Rogue space planting	Rice brown plant hopper
7.	Plant density	Rice brown plant hopper
8.	Earthing up	Sugarcane whitefly
9.	Detrashing	Sugarcane whitefly
10.	Destruction of weed hosts	Citrus fruit sucking moth
11.	Destruction of alternate host	Cotton whitefly
12.	Flooding	Rice armyworm
13.	Trash mulching	Sugarcane early shoot borer
14.	Pruning/topping	Rice stem borer
15.	Intercropping	Sorghum stem borer
16.	Trap cropping	Diamond back moth
17.	Water management	Brown plant hopper
18.	Judicious application of fertilizers	Rice leaf folder
19.	Timely harvesting	Sweet potato weevil

In rice, MDU 3 against gall midge, PY 3, Co 42 and ADT 44 against brown plant hopper; in sorghum SH 15R against shoot fly; in groundnut robust 33-1, Kadiri-3, ICGS 86031 against thrips and bud necrosis disease and in cotton (tolerant varieties) JGJ 14545, LK 861, Supriya and Kanchana against whitefly can be grown in endemic areas.

IV. Botanical Insecticides

Use of botanical insecticides for pest management is increasing owing to increasing consciousness about food safety. Around 2400 plant species with pesticidal properties have been reported.

Among the plant products, neem oil 2 per cent, neem seed kernel extract (NSKE) 5 per cent with teepol 0.05 per cent and commercial formulations like Azadirachtin are quite effective against major pests of rice, sucking pests of cotton, vegetables *etc.* Neem coated urea is also helpful in slow release of nutrients. The plant products are also safer to non-target organisms and entomophages.

V. Behavioural Pest Management

a. Pheromones in Pest Management

Pheromones are the chemical substances released by insects which attract other individuals of the same species. Sex pheromones are being used in pest management for monitoring, mating disruption and mass trapping. Pheromones are species specific and their safety to beneficial organisms makes them ideal components of integrated pest management systems. Of the various pheromones, the sex and aggregation pheromones are used for pest management in India. Synthetic sex pheromones/aggregation pheromones are placed in traps to attract males. The rubberized septa, containing the pheromone lure are kept in traps designed specially for this purpose and used in insect monitoring/mass trapping programmes.

Sex/aggregation pheromones of the following insects are commercially available in market for pest management (Table 6.4).

Table 6.4: Commercially available Sex/Alarm Pheromones for Insect Pest Management

Sl.No.	*Insect Pest*
	Sex Pheromones
1	American bollworm (*Helicoverpa armigera*)
2	Spotted bollworm (*Earias vitella*)
3	Tobacco cutworm (*Spodoptera litura*)
4	Rice yellow stem borer (*Scirpophaga incertulas*)
5	Spiny bollworm (*Earias insulana*)
6	Pink bollworm (*Pectinophora gossypiella*)
7	Sugarcane shoot borer (*Chilo infuscatellus*)
8	Diamond back moth (*Plutella xylostella*)
9	Mango fruit fly (*Bactrocera dorsalis*)
10	Melon fruit fly (*Bactrocera cucurbitae*)
11	Tomato pinworm (*Tuta absoluta*)
	Aggregation Pheromone
1	Coconut rhinoceros beetle (*Oryctes rhinoceros*)
2	Coconut red palm weevil (*Rhyncophorus ferrugineus*)

b. Light Trap

The behaviour of certain species of insects being attracted to light could be advantageously used in their management. The light traps could be used both for monitoring and for mass trapping. Rice stem borer and brown plant are attracted towards yellow light, while the rice leaf folder and green leaf hoppers towards green light source.

c. Yellow Sticky Trap

Cotton whitefly, aphids, thrips prefer yellow colour. Yellow colour is painted on tin boxes and sticky material like castor oil/vaseline is smeared on the surface and kept @ 5ac^{-1}. These insects are attracted to yellow colour and trapped on the sticky material.

d. Bait Trap

Attractants are used in traps to attract the insect and kill some of the pests. (eg.) Fishmeal trap: This trap is used against sorghum shoot fly. Moistened fish meal is kept in polythene bag or plastic container inside a container along with cotton soaked with insecticide (dichlorvos) @ 5 $ac^{-1}$to kill the attracted flies.

e. Probe Trap

Probe trap is used by keeping them under grain surface to trap stored product insect.

VI. Physical Methods

Activated clay at 1 per cent or vegetable oil at 1 per cent effectively reduces the pulse beetle damage. Heat treatment of sorghum seeds using solar drier for 60 seconds reduce the rice weevil, *Sitophilus oryzae* and red flour beetle, *Tribolium castaneum* damage. Hot water treatment at 50 to 55°C for 15 minutes checks the rice white tip nematode. Use of flame throwers checks the population of locusts and burning torches checks hairy caterpillars. Cold storage of fruits and vegetables at 1 to 2°C for 12 to 20 days kills fruit flies.

VII. Transgenic Crops

Plants engineered with the crystal protein gene are being commercialized in tomato, cotton and tobacco. Monsanto commercialized the Bt transgenic cotton (Bollgard), corn (Yield Gard) and potato (New leaf). Novartis has commercialized a transgenic corn (Maximiser).

VIII. Insecticides

Use of insecticides in IPM is not an ecological sin, but it requires several changes in the way the insecticides are used, quantities applied and the nature of insecticides selected. In fact, insecticides are one of the components of most IPM systems because of their convenience, simplicity, effectiveness, flexibility and economy. But their use must be judicious and integrated with other IPM components. IPM discourages the use of broad spectrum and persistent insecticides and encourages selective insecticides.

1. Pesticide should be applied only based on the need, *i.e.* if pest reaches ETL
2. It should be judiciously combined with other components of IPM and pesticides should be used as last resort
3. When pest population approaches ETL, insecticides are the only means of preventing economic damage
4. Pesticides which are cost effective (High benefic/cost ratio) and safe (High benefit/risk ratio) should be used in IPM

Many new insecticide molecules with less persistence like chitin synthesis inhibitors, neonicotinoids, avermectins, juvenile hormone analogues, milbemycins, spinosyns *etc.* are also available commercially which can be effectively integrated with other IPM approaches.

Table 6.5: IPM Practices for Major Pests of Agricultural Crops

Rice	☆ Remove/destroy stubbles after harvest ☆ Keep the fields free from weeds ☆ Trim field bunds ☆ Provide effective drainage, if required ☆ Avoid use of excessive 'N' fertilizers ☆ Avoid close planting, especially in BPH and leaf folder prone areas and seasons ☆ Leave 30 cm space at every 2.5 M ☆ Use irrigation water judiciously ☆ Use light traps to monitor pest incidence ☆ Remove egg masses of stem borer ☆ In BPH prone areas/seasons, avoid use of synthetic pyrethroids, methyl parathion and quinalphos and use recommended chemical at recommended doses ☆ Use insecticides based on ETLs
Stem borer *Scirpophaga incertulas*	☆ Release egg parasitoid, *Trichogramma japonicum* @ 40,000/ac at 25 per cent of the ETL for insecticides (2 egg masses/m^2) or ☆ Spray neem seed kernel extract 5 per cent at 25 per cent of the ETL for insecticides (2 egg masses/m^2) ☆ If needed, spray azadirachtin 0.03 per cent @ 400 ml/ac or cartap hydrochloride 50 per cent SP @ 400 g/ac or chlorpyriphos 20 per cent EC 500 ml/ac or fipronil 5 per cent SC 400ml/ac
Brown plant hopper *Nilaparvata lugens*	☆ Avoid excessive use of nitrogen ☆ Control irrigation by intermittent draining ☆ Set up light traps during night or yellow pan traps during day time ☆ Drain the water before use of insecticides and direct the spray towards the base of the plants ☆ Spray neem oil 3 per cent @ 6 lit/ac with soap oil ☆ If needed, spray dichlorvos 76 per cent SC 200 ml/ac or buprofezin 25 per cent SC @ 325ml/ac or fipronil 5 per cent SC 400ml/ac or imidacloprid 17.8 per cent SL 40 ml/ac can be used
Pulses	
Pod borers ☆ Blister beetle Mylabris spp. ☆ Spotted pod borer, *Maruca vitrata* ☆ Plume moth, *Exelastis atomosa*	☆ For pod borers, raise one row of sunflower as intercrop for every 9 rows of pigeon pea and plant maize as border crop ☆ Keep pheromone traps for *Helicoverpa armigera* @ 5/ac ☆ Erect bird perches 20/ac ☆ Mechanical collection of grown up larva and blister beetle ☆ Apply HaNPV 1.2 x10^{12} POB/ac in 0.1 per cent teepol

☆ Grampod borer *Helicoverpa armigera* ☆ Pod fly, *Melanagromyza obtusa* ☆ Pod bug, *Clavigralla gibbosa*	☆ If needed, spray azardirachtin 0.03 per cent @ 400 ml/ac or Emamectin benzoate 5 per cent SG 88 g/ac or Indoxacarb 15.8 per cent SC 133 ml/ac or NSKE 5 per cent twice followed by triazophos 0.05 per cent or Neem oil 2 per cent
Red hairy caterpillar *Amsacta albistriga*	☆ Dig out and destroy the pupae from the field bunds and shady places ☆ Set up 3 to 4 light traps and bonfires immediately after receipt of rains, after sowing in the rainfed season to attract and kill the moths and also to know brood emergence ☆ Collect and destroy gregarious, early instar larvae on lace-like leaves of intercrops such as redgram and cowpea ☆ Collect and destroy egg masses in the cropped area ☆ Avoid migration of larvae by digging a trench 30 cm deep and 25 cm wide with perpendicular sides around the infested fields ☆ Use virus suspension obtained from 300 medium sized larvae for spraying one acre along with a sticker 100 ml or Triton in 140 l of water. Use potable water for mixing and spray in the evening hours ☆ If needed, spray phosalone 35 EC 300 ml/ac or dichlorvos 76 EC 250 ml/ac
Tobacco cut worm *Spodoptera litura*	☆ Grow castor as border or intercrop in groundnut fields to serve as indicator or trap crop ☆ Monitor the emergence of adult moths by setting up light and pheromone traps ☆ Collect egg masses and destroy ☆ Collect the gregarious larvae and destroy them as soon as the early symptoms of lace-like leaves appear on castor, cowpea and groundnut ☆ Apply nuclear polyhedrosis virus (NPV) 0.6×10^{12} POBs/ac with crude sugar 1.0 kg/ac and Teepol 100 ml/ac ☆ If needed spray dichlorvos 76 WSC 300 ml/ac or Diflubenzuron 25 WP 150g/ac
Cotton	☆ Remove the cotton crop and dispose of the crop residues as soon as harvest is over ☆ Avoid stacking of stalks in the field ☆ Avoid ratoon and double cotton crop ☆ Adopt proper crop rotation ☆ Use optimum irrigation and fertilizers ☆ Grow one variety throughout the village as far as possible
American bollworm *Helicoverpa armigera*	**Monitoring** ☆ Pest monitoring through light traps, pheromone traps and *in situ* assessments by roving and fixed plot surveys has to be intensified at farm, village, block, regional and State levels ☆ For management, an action threshold of one egg per plant or 1 larva/plant may be adopted

	Cultural practices
	✰ Synchronized sowing of cotton preferably with short duration varieties in each cotton ecosystem
	✰ Avoid continuous cropping of cotton both during winter and summer seasons in the same area as well as ratooning
	✰ Avoid monocropping
	Chemical control
	✰ If needed, spray chlorpyriphos 20 EC 800 ml/ac (or) phosalone 35 EC 600 ml/ac
Pink bollworm	✰ Use pheromone trap to monitor the adult moth activity
Pectinophora gossypiella	✰ Three weekly releases of egg parasitoid *Trichogrammatoidea bactrae* @ 40,000/ac per release coinciding with the incidence of the pest
	✰ If needed, spray phosalone 35 EC 600 ml/ac
	Sugarcane
EarlyShoot borer, *Chilo infuscatellus*	✰ Early season planting (Dec-Jan) should be done
	✰ Trash mulching on ridges on 3DAP
	✰ Intercropping with green gram, black gram, daincha effectively checks shoot borer
	✰ Installation of early shoot borer sex pheromone traps @ 4/ac for monitoring of pest density
	✰ Spray Granulosis Virus (GV) at 0.6 x 10^{12} PIB/ac twice on 35 and 50 days after planting (DAP) or release 50 gravid females of *Sturmiopsis inferens*/ac on 30 and 45 DAP during evening hours
	✰ If needed, spray Monocrotophos 36 WSC 400 ml/ac

Conclusion

By properly integrating all these methods, insect pest on different crops can be managed in a sustainable manner without causing any ill-effects to human beings, non-target beneficial arthropods as well as the environment.

Chapter 7

Integrated Disease Management of Major Crop Plants

D. Alice and M. Karthikeyan

Department of Plant Pathology,
Tamil Nadu Agricultural University,
Coimbatore – 641 003, Tamil Nadu

A disease is a condition in a plant that affects the plants normal functioning or development. Three conditions are essential for the establishment of a disease. They are pathogen, susceptible host and conducive environmental factors.Integrated disease management is the practice of using a range of measures to prevent and manage diseases in crops. During the cropping cycle, regular crop monitoring is used to decide if and what action is needed to tackle the diseases.

Common IPM Strategies for Crops

☆ Use clean seed and vegetative propagating material. Some disease-causing organisms are seed-borne and others are associated with seeds. Use only disease free seedlings.

☆ Select cultivars those are resistant or tolerant to pathogens.

☆ Destroy or remove infected crop residues, Crop residues harbour disease organisms and hence can be burned, composted or buried.

☆ Rogue infected plants up to 45 days of sowing.

☆ Rotate crops to avoid or reduce the build-up of disease organisms in a field. Note that some plants have shown to have a suppressive effect on diseases.

- ✰ Land should be free of volunteer plants.Remove collateral weed hosts from bunds and channels
- ✰ Provide plants with good nutrition as properly nourished plant can withstand or tolerate the attack of plant pathogens. Avoid excess nitrogen, apply N in three split doses.Apply potash fertilizers to protect crop from diseases
- ✰ Avoid injuring or bruising plants because many pathogens can enter a plant through an injury or wound.
- ✰ Use proper spacing to allow air movement between plants and the leaf wetness should be reduced.
- ✰ Schedule timing and duration of irrigation to satisfy the crop requirements without over-watering. Too much or too less water can be detrimental.
- ✰ Select the appropriate control method — biocontrol or chemicals (fungicides, bactericides and fumigants).Timely application will save the crop from diseases.

Bio-control Practices for all Crops

- ✰ Seed treatment: *Pseudomonas fluorescens* @ 10g/kg (or) *Trichoderma viride* @ 4g/kg of seed.
- ✰ Mix *Trichoderma viride* or *P. fluorescens* @ 2.5kg/ha with 50 kg of well decomposed Farmyard Manure and apply to the field. The fortification of biocontrol agents with Farmyard manure can be done 10 days before application.
- ✰ Effective against root rot, wilt and all soil borne diseases.

For Rice

- ✰ Treat the seeds with talc based formulation of *P. fluorescens* (Pf1) @ 10g/kg of seed and soak in 1lit of water overnight. Decant the excess water and allow to sprout the seeds for 24h and then sow.
- ✰ Seedling dip with *Pseudomonas fluorescens*: Stagnate water to a depth of 2.5cm over an area of 25m^2 in the main field. Sprinkle 2.5 kg of the talc based formulation of *Pseudomonas fluorescens* (Pf1) and mix with stagnated water. The seedlings pulled out from the nursery are to be soaked for 30 min. in the stagnated water and then transplanted.
- ✰ Foliar spray with *P. fluorescens* 0.2 per cent is effective against rice diseases.

Cereals

Rice

1. Blast: *Pyricularia grisea* (*Magnoporthe grisea*)

- ✰ All aerial parts of the plant (leaf blade, node neck of the panicle rachis and also glumes) affected both in the nursery and main field.

- ☆ Minute brown specks appear on leaves which become spindle – shaped with dark brown margin and grayish center with yellow halo. **(Leaf blast)**
- ☆ The nodes turn black and rot. **(Node blast)**
- ☆ Neck region rots and the ear head becomes black and the panicle break (neck rot/**neck blast**) Complete chaffiness of grains or partial filling of grains occur.

2. Brown Spot: *Helminthsporium oryzae* (*Drechslera oryzae*)/*Cochliobolus miyabeonus*

- ☆ Brown spots on leaves and grains.
- ☆ Initially minute dark brown to black dots later become oval or rectangular.
- ☆ Discolored seeds are shriveled.

3. Sheath Rot: *Sarocladium oryzae*

- ☆ **Stage of infection:** Boot leaf stage.
- ☆ Flag leaf exhibits oblong or irregular, long spots with grey center and brown margin.
- ☆ Complete choking of ear head or partial emergence of the panicle
- ☆ Un emerged panicles rot. The glumes are also discoloured.

4. Sheath Blight: *Rhizoctonia solani*

- ☆ **Stage of infection:** Tillering stage.
- ☆ Oval greenish grey spots on leaf sheaths near the water level.
- ☆ Lesions are large, oblong or irregularly elongated with grayish white center and brown margin. In advanced stages brown spherical sclerotia are formed. Also infects panicle thus becoming seed borne.

5. Bacterial Leaf Blight: *Xanthomonas oryzae* pv *oryzae*

- ☆ Blighting of seedling occurs in nursery. In main field, "Kresek' phase *i.e.* death of seedling is usually observed one or two weeks after transplanting.
- ☆ Dull greenish or yellowish water soaked spots seen near the leaf tip or margins. Spots enlarge turn yellow extends along one or both margins towards midrib and downwards (wavy margin) leaving central green healthy area. This is called Flag symptom.
- ☆ Several lesions coalesce to form straw coloured lesions or blighted portions. Droplets of pale, amber coloured bacterial ooze seen on the affected portion.

6. Rice Tungro Virus: RTV

- ☆ Plants much stunted and number of tillers reduced. Leaves become yellow to deep orange. Young leaves show mottling or pale green stripes.
- ☆ Results in poorly developed root system and no panicle formation.

- ☆ The virus is transmitted by green leaf hoppers. *Nephotettix virescens, N. nigropictus* and Zigzag hopper *Recilia dorsalis*.

IDM Practices for Rice Diseases

- ☆ **Dry seed treatment:** Thiram or captan or carboxin or carbendazim at 2 g/kg of seeds. Treat the seeds at least 24 hours prior to soaking for sprouting. The treated seeds can be stored for 30 days without any loss in viability.
- ☆ **Wet seed treatment:** Carbendazim or Tricyclozole at 2 g/lit of water for 1 kg of seed. Soak the seeds in the solution for 2 h. Drain the solution, sprout the seeds and sow in the nursery bed. This wet seed treatment gives protection to the seedlings up to 40 days from seedling disease such as blast.
- ☆ For blast disease management spray Carbendazim 50WP @ 500g/ha (or) Tricyclozole 75 WP @ 500g/ha.
- ☆ For sheath rot disease management apply Gypsum @ 500 kg/ha at two equal splits once basally and another at active tillering stage. Spray any one of the chemical, Carbendazim @ 500g/ha (or) Hexaconazole 75 per cent WG @ 100 mg/lit. 1st spray at the time of disease appearance and 2nd spray 15 days later.
- ☆ For the management of Rice grain discoloration spray Carbendazim + Thiram + Mancozeb (1:1:1) 0.2 per cent at 50 per cent flowering stage.
- ☆ For the management of Bacterial blight, spray fresh cow dung extract 20 per cent twice (starting from initial appearance of the disease and another at fortnightly interval) or two sprays of Copper hydroxide 77 WP@1.25 kg/ha 30 DAP and 45 DAP.
- ☆ For the management of False smut disease, two sprayings of Propiconazole 25 EC @ 500ml/ha (or) Copper hydroxide 77 WP @ 1.25 kg/ha at boot leaf and 50 per cent flowering stages.

Maize

1. Common Smut

- ☆ Galls on ears, auxiliary buds, tassels stalks and rarely on leaves. Embryonic tissues of the plants are affected. Galls covered with white membrane, later rupture and expose the black mass of powdery spores.
- ☆ Galls on young seedlings results in stunting or death of plants.

2. Head Smut: *Sporisorium reilianum* (*Sphacelotheca reiliana*)

- ☆ Smut sori formed on tassel and ear. Flower parts are covered by a black powdery mass of spores.
- ☆ Inside sori spores mixed with network of fibrous tissue. Floral bracts grow out into a leaf structures or into small shoots.

3. Downy Mildews: *Peronosclerospora sorghi*

- ☆ Interveinal narrow chlorotic yellowish stripes seen on leaves. Stripes become brown or straw coloured and necrotic. Early infection causes no seed development and plants die prematurely.

4. Charcoal Rot: *Macrophomina phaseolina*

- ☆ Basal internodes of the infected plants become soft, and weak resulting in lodging. If infected stem split open, small black sclerotia seen within the mass of vascular fibers. Hence the disease is called as charcoal rot.

Integrated Disease management:

- ☆ For downy mildew, spray metalaxyl @ 1000g/ha and for leaf blight diseases spray mancozeb 2g/lit 20DAS. Grow multiple disease resistant hybrid CoH6.

Millets

Sorghum

1. Grain Smut: *Sporisorium sorghi* (*Sphacelotheca sorghi*)

- ☆ Majority of the grains affected. Individual grains are replaced by oval dirty grey sac like structure called smut sorus. Sori are covered with a tough white or creamy to light brown membrane which often remain unbroken up to threshing. The sorus is completely filled with dark brown spore powder.

2. Loose Smut: *Sporisorium cruenta* (*Sphacelotheca cruenta*)

- ☆ Plants stunted, produce thinner stalks, more tillers and earlier flowering than healthy ones. Affected inflorescences are looser and bushier with hypertrophied glumes. All florets are affected and converted into smut sori.
- ☆ The membrane covering the sorus ruptures exposing the smut spores.

3. Ergot or Sugary Disease: *Sphacelia sorghi*

- ☆ Individual spike lets in the ear affected. Creamy sticky liquid ooze out from the affected spike lets. Ooze blackened due to saprophytic fungi. Long straight or curved, brownish, hard sclerotia produced from the affected spike lets.

4. Head Mould: *Fusarium*, *Curvularia*, *Alternaria* and *Aspergillus* sp.

- ☆ Severe grain mould occur during rainy seasons and are either pink or black in colour.

Integrated Disease Management

- Seed treatment: Treat the seeds 24h prior to sowing with Carbendazim or Captan or Thiram 2g/kg of seed or Metalaxyl 6g/kg of seed.
- For rust disease, Spray Mancozeb at 1kg/ha and repeat fungicidal application after 10 days.
- Ergot or Sugary disease/head mould: Sowing period to be adjusted so as to prevent heading during rainy season and severe winter. Spray Mancozeb – 1000g/ha or Propiconazole 500ml/ha at emergence of earhead (5 - 10 per cent flowering stage) followed by a spray at 50 per cent flowering and repeat the spray after a week if necessary.
- Downy Mildew: Spray any one of the fungicides like Metalaxyl + Mancozeb 500 g or Mancozeb 1000g/ha after noticing the symptoms of foliar diseases, for both transplanted and direct sown crops.

Redgram

1. Wilt: *Fusarium oxysporum* f.sp. *udum*

- Gradual or sudden wilting from bottom to top. Leaves initially pale, loose their turgidity and droop (withering).
- Entire plants wilt or die within few days. Disease occur in patches
- Vascular browning seen indicating xylem plugging with mycelia.

2. Sterility Mosaic

- Plants stunted due to shortening of internodes and look chlorotic
- Auxiliary buds grow, crowded at the top giving bushy appearance.
- Leaves become small and crinkled with mild mottle to severe mosaic.
- Ring spot symptom seen. Vector: Mite *Aceria cajani*

3. Powdery Mildew: *Leveillula taurica*

- Powdery patches on the lower surface corresponding yellow patches on upper surface. Premature defoliation.

4. Leaf Spot: *Cercospora cansescens*

- Light brown spots bound by veins on under surface of the leaves.
- Premature defoliation.

Integrated Disease Management

- For sterility mosaic virus management, rogue out the infected plants in the early stages of growth. Spray Fenazaquin @ 1 ml/lit on 45 and 60 DAS as prophylactic spray.

Black Gram, Green Gram

1. Dry Root or Charcoal Rot: Sclerotial Stage *Rhizoctonia* batataticola, Pycnidial stage: *Macrophomina phaseolina*

- ☆ Yellowing and dropping of leaves occurs in patches, plants wilt and dry suddenly. Infected plants easily pulled out.
- ☆ Grayish black sunken lesion leading to bark shredding Dark brown sclerotia seen on roots and black minute pycnidia on stem. Rotting of roots.

2. Powdery Mildew: *Erysiphe polygoni*

- ☆ White powdery growth on the upper surface of leaves.
- ☆ In advanced stages powdery growth turns brown and leaves drop off prematurely.

3. Leaf Spot: *Cercospora canescens*

- ☆ Well defined, reddish brown spots with grey center bounded by veins.
- ☆ Shot hole symptoms seen.

4. Yellow Mosaic Disease

- ☆ Small irregular yellow patches in between veins giving mosaic mottling.
- ☆ Yellowing covers the entire leaf.
- ☆ Brown necrotic spots are also seen.
- ☆ **Vector:** white fly *Bemisia tabaci*

Integrated Disease Management in Blackgram and Greengram

- ☆ For yellow mosaic disease management, growing resistant varieties in black gram such as VBN 4, VBN 6 and VBN 7. Seed treatment with Dimethoate (or) Imidacloprid @ 5 ml/kg. Growing 7 rows of sorghum as border crop. Installation of yellow sticky traps 12 Nos./ha. Rogue out the yellow mosaic disease infected plants up to 45 days. Foliar spray of notchi leaf extract 10 per cent at 30 DAS or neem formulation 3 ml/lit or methyl demeton 25 EC 500 ml/ha or dimethoate 30 EC 500 ml/ha or thiamethoxam 75 WS 1g/3 lit and repeat after 15 days, if necessary.
- ☆ For Powdery Mildew disease management, spray NSKE 5 per cent or Neem oil 3 per cent twice at 10 days interval from initial disease appearance. Spray Eucalyptus leaf extract 10 per cent at initiation of the disease and 10 days later. Spray Carbendazim 500 g or wettable sulphur 1500g/ha or Propiconazole 500 ml/ha at initiation of the disease and 10 days later.
- ☆ For leaf spot disease management, spray Mancozeb 1000g or Carbendazim 500 g/ha at initiation of the disease and 10 days later.
- ☆ For root rot management, basal application of neem cake @ 150 kg/ha and spot drenching of Carbendazim @ 1 gm/lit.

- ☆ Root rot-stem fly complex management, seed treatment with *Beauveria bassiana* + *Pseudomonas fluorescens* @ 5 g each/kg of seed.

Oilseeds

Groundnut

1. Early Leaf Spot: Tikka Leaf Spot: *Cercospora arachidicola/Mycospherella arachidis*

- ☆ Affects all aerial parts – mostly on leaf. Occurs early in crop season – 3 to 4 weeks after sowing Irregular reddish brown spots with yellow halo Leaves shed prematurely and yield reduced.

2. Late Leaf Spot: *Cercospora personata/Mycospherella berkeleyii*

- ☆ Spots appear 5 to 7 weeks after sowing Dark brown circular spots without yellow halo.
- ☆ The lower surface lesions carbon black. Severe infection –leaf shed prematurely and yield reduced.

3. Rust: *Puccinia arachidis*

- ☆ Infection seen on 6 weeks old crop. Brick red pustules on lower surface of leaf.
- ☆ Upper surface necrotic brown spots, Pustules seen on leaf and stem.
- ☆ Dark coloured teliosori appear which produces teliospores.
- ☆ Leaves dry,drop prematurely, seeds small and shriveled.

4. Peanut Spotted Wilt/Bud Necrosis/Groundnut Ring Mosaic/TSWV

- ☆ Ring spot, mottling and rugosity of leaves. Leaflets size reduced and distorted.
- ☆ Necrotic spots on stem and leaf. Bud necrosis.
- ☆ Pods wrinkled with black lesions on testa.

Integrated Disease Management

- ☆ For rust disease and tikka leaf spot, spray Mancozeb 1000g/ha or Chlorothalonil 1000g/ha or Wettable sulphur 2500g/ha or Tridemorph 500 ml/ha If necessary, repeat the spray 15 days later.
- ☆ Groundnut Bud Necrosis management, adopt a close spacing of 15 x 15 cm. Remove infected plants up to 6 weeks after sowing and spray Monocrotophos 36 WSC 500 ml/ha, 30 days after sowing either alone or in combination with antiviral principles. Antiviral principles from sorghum or coconut leaves. AVP are extracted as follows: Sorghum or coconut leaves collected, dried, cut into small bits and powdered to one kg of leaf powder two litres of water is added and heated to 60°C for one

hour. It is then filtered through muslin cloth and diluted to 10 litres and sprayed. To cover one ha 500 litre of fluid will be required. Two sprays at 10 and 20 days after sowing will be needed.

Integrated Disease Management

- ☆ For the management of Alternaria blight and Cercospora leaf spot, spray Mancozeb 1000g/ha.
- ☆ For the management of Phyllody, remove and destroy infected plants. For the management of vector, spray Monocrotophos 36 or Dimethoate 30 EC 500 ml/ha combined with intercropping of Sesamum + Redgram (6 : 1) DAS.

Cotton

1. Wilt: *Fusarium oxysporum* f.sp *vasinfectum*

- ☆ First symptoms on young seedlings are yellowing and browning of cotyledons, followed by brown ring on the petiole.
- ☆ Finally wilting and drying of the seedling occur. Infection at later stages includes loss of turgidity, yellowing, drooping and wilting, starting from older leaves. Browning or blackening of vascular tissues occur on the stem and spreads upwards and downwards. Infected plants stunted with fewer bolls.

2. Boll Rot: *Fusarium moniliforme, Aspergillus flavus*

- ☆ Brown or black dots covering entire bolls. Botting may be internal or external. Bolls do not open and fall prematurely.

3. Leaf Blight: *Alternaria macrospora*

- ☆ It affects all stages of crop. Initially, produces small, brown irregular to round spots with a central necrotic lesions which on coalesce form large blighted areas.
- ☆ The affected leaves brittle and fall off. Symptoms are also in stems, bracts and bolls in severe cases.

4. Bacterial Blight: *Xanthomonas campestris* p.v *malvasearum*

- ☆ Water soaked, circular or irregular lesions on cotyledons which spread to petiole and stem and finally withering and death of seedling known as Seedling blight.
- ☆ The infection of veins and veinlets shows blackening with crinkled and twisted leaves and bacterial oozing.
- ☆ Black lesions on stem and branches, premature drooping off of the leaves resulting in die back known as Black arm. It also affects the bolls causing boll rot.

Integrated Disease Management

- ✰ For bacterial leaf blight disease management, avoid stacking of infected plants. Spray Streptomycin sulphate + Tetracycline mixture 100g + Copper oxychloride 1250g/ha. Repeat spraying at 10 days interval twice or thrice if drizzling continues.
- ✰ For Alternaria leaf spot, Grey Mildew and boll rot management, spray Copper Oxychloride 1250g or Mancozeb 1000g or Chlorothalonil 500g/ha at 60, 90 and 120 days after sowing.

Sugarcane

1. Red Rot: *Colletotrichum palcatum*

- ✰ Red rot in the internodes of stalk. Reddening of internal tissues at right angles.
- ✰ Cross wise white patches. Acidic sour smell.
- ✰ Minute black velvetty fruiting bodies seen.
- ✰ Blood red lesions with straw coloured margin seen on leaves.

2. Smut: *Ustilago scitaminea*

- ✰ Culmiculous smut.
- ✰ Central shoot converted into long whip like dusty black structure.
- ✰ Whip covered by white papery membrane, in maturity membrane ruptures and liberates smut spores (Teliospores).
- ✰ Mummified arrows.

Integrated Disease Management

- ✰ For the management of red rot, sett rot and smut disease, adopt sett treatment with Carbendazim before planting (Carbendazim 50 WP @ 0.05 per cent or Carbendazim 25 DS @ 0.1 per cent along with 1.0 per cent Urea for 5 minutes).
- ✰ For red rot disease management, growing of recommended resistant and moderately resistant varieties *viz.*, Co 86249, CoSi 95071, CoG 93076, CoC 22, CoSi 6 and CoG 5. The irrigation interval in a red rot affected field must be lengthened. Once in 15 days during tillering, growth phases and once in 25 days during maturity phase which restricts the spread. Removal of the affected clumps at an early stage and soil drenching with 0.1 per cent Carbandazim 50 WP or 0.25 per cent lime. The trash of red rot affected field after harvest may be uniformly spread and burnt. The red rot affected field must be rotated with rice for one season and other crops for two seasons.

- ☆ Coc 22. Spray dimethoate @ 0.1 per cent to control insect vector. Avoid ratooning if GSD incidence is more than 15 per cent in the plant crop.
- ☆ Spray Mancozeb 2.0 kg or Carbendazim 500 g/ha for the management of fungal diseases.

Chapter 8

Management Techniques of Polluted Soil and Water for Complex, Diversified and Risk Prone Agriculture

P.T. Ramesh, P. Kalaiselvi, K. Suganya, S. Paul Sebastian, M.P. Sugumaran, R. Jayashree and K. Boomiraj

Department of Environmental Sciences,
Tamil Nadu Agricultural University,
Coimbatore – 641 003, Tamil Nadu

The land and water resources of the country are exploited continuously due to increasing population, rapid urbanisation and industrialisation. Failure of monsoon and changes in climate adds up to resource depletion of water resources, even in catchment areas and river basins. Considering the current water availability and water use pattern, it is imperative that India should concentrate on wastewater recycling in future. India, with a share of 2.45 per cent of totalworld's land availability and 4 per cent of water resources, has to cater to the needs of its 120 million people, which is 16 per cent of that of the world. It is estimated that India consumes on an average of 1123 billion cubic meter annually, of which 85 per cent (688 million cubic meter) is consumed for agriculture alone. This estimate is expected to go up to 1072 million cubic meter by 2050 (CGWB, 2011).

While majority of this water is used for agriculture, remaining is released into rivers, ponds, lands and oceans partly treated or untreated. This cause serious pollution of land, water and also contaminate food produced.

An Introduction to Problem Soils

Land degradationis a temporary or permanent lowering of the productive capacity of land, as defined by UNEP. It thus covers the various forms of soil degradation including those by human impacts. Sodicity and salinisation are two major issues arising out of effluent discharge in soil. Tanneries, textile and dye industries, pulp and paper manufacturing industries, distillery industries are the major pollution contributing industries, as illustrated in Table 8.1. Though salinization and sodicity in soils are locally occurring phenomena, secondary soil salinization and sodicity arises due to discharge of untreated effluent from these industries. High levels of salinity and sodicity leads to build up of free salts and codification (alkalization), change in soil physical properties, immobilisation of nutrients, retarded plant growth and reduced crop productivity. The impact of pollution is also influenced by the soil property, topography of the field and cultural practices.

Table 8.1: Industrial Effluents Causing Salinity and Sodicity in Soil

Parameters	*Tannery*	*Pulp and Paper*	*Distillery*
pH	6 -12	8.25	5.6 – 8.4
EC (dSm^{-1})	11 – 55	1.90	0.4 – 1.5
ESP	21.6	34.78	6.9 – 42.0
Soil category	Saline sodic	Saline sodic	Saline sodic

An Introduction on Wastewater

Wastewater, in general, refers to the water that is discharged out from various processes in processing or production units and households. Depending on the source and type of raw material used, the constituents of wastewater differ. Apart from specific pollutants like heavy metals, toxic organics from industries, pH and salinity of the wastewater is of great concern with wastewater. However, most wastewaters contain considerable quantum of plant nutrients and hence such non- toxic wastewaters are recommended for irrigation after proper treatment. Continuous use of treated effluents for irrigation also adds to groundwater pollution by infiltration and percolation.

The Central Pollution Control Board (CPCB) has prescribed standards for irrigation of treated wastewater. The standards for key parameters are furnished in Table 8.2.

Management of such polluted soil and water requires an integrated approach involving proper treatment of wastewater, addition of suitable soil amendments, selection of crop varieties and irrigation frequency. Remedies evolved through various research programmes for management of polluted soils and water due to major industries of Tamil Nadu is briefly discussed.

Table 8.2: General Standards Prescribed by CPCB for Treated Wastewater Irrigation

Quality Parameter	*For Discharge into Land for Irrigation*
pH	5.5 – 9.0
Temperature (ºC)	Not to exceed 5ºC above receiving water temperature
BOD (mg L^{-1})	100
Chlorides (mg L^{-1})	600
Sulphates (mg L^{-1})	1000
Oil and grease (mg L^{-1})	10
Arsenic (mg L^{-1})	0.2

I. Impact of Tannery Effluent on Soil and Water Quality

Leather industry is an important foreign exchange earning industry. There are 2091 tanneries in India to process these skins, having an annual processing capacity of around 7 lakh tones of hides and skins. Off these, 45 per cent are in Tamil Nadu, 26 per cent in West Bengal, 18 per cent in Uttar Pradesh, accounting for more than 85 per cent of tannery units in the country. Major tanning clusters in the country are Chennai, Ambur, Ranipet, Kolkata, Kanpur and Jalandhar.

Characteristics of Tannery Wastewater

Leather processing involves curing, soaking, liming and unhairing, deliming, bating, pickling, East India (also called EI or vegetable tanning) tanning or chrome tanning and washing. While EI tanning does not use chromium (Cr) based salts and depend upon natural tanning agents, chromium is added instead of natural agents in chrome tanning. This makes both effluents differ in their property.

Properties of Vegetable Tanning Effluent

The spent vegetable tan liquor is very strong in colour and composition with an estimated generation of about 200 to 400 litres per quintal of hide or skin processed (Kaul *et al.*, 2005). It is effluent is deep or dark brown in colour with a characteristic offensive odour. The pH of the effluent ranges from 5.0 to 6.3 and the BOD of the effluent is between 3300 and 12000 mg L^{-1}. Table 8.3, depicts the characteristics of treated vegetable tan effluent.

However, the effluent composition varies with the type of raw material used and the process employed. Bosnic *et al.* (2003) have attributed much higher pollution load in vegetable tan effluents. They reported high BOD (10,000-20,000 mg L^{-1}), COD (20,000-35,000 mg L^{-1}) and suspended solids (10,000-15,000 mg L^{-1}). Further, vegetable tannins are hard biodegradable materials. Therefore, waste bearing vegetable tannins degrade, although slowly.

The one advantage with this method is that it does not need the prior preparation stage of pickling and therefore the contribution to pollution load from sulphate salts is lower.

Table 8.3: Characteristics of Vegetable Tanning Effluent

Parameter	*Value*
pH	8.7 – 9.5
Suspended solids (mg L^{-1})	3000 – 5600
Total Dissolved Solids (mg L^{-1})	8500 – 19680
Biochemical Oxygen Demand (mg L^{-1})	2300 – 2650
Chemical Oxygen Demand (mg L^{-1})	5320 – 7160
Sulphide (mg L^{-1})	75 – 90
Total chromium (mg L^{-1})	8 – 22
Oil and grease (mg L^{-1})	17 - 43

Source: Sankar, 2001.

Properties of Chrome Tanning Effluent

Chrome tanning process uses dichromate, sodium chromate and chromium sulphate salts having 50 per cent basicity as raw material. This adds tremendous chromium wastes in wastewater. The quantity of Cr compounds added to leather units alone is a whooping 69,000 tons; 25 to 39 per cent has been spent in wastewater (Iyer and Mastorakis, 2006). Not only chromium salts, but large quantities of sodium chloride, NaCl; calcium hydroxide, Ca $(OH)_2$; sodium sulphide, Na_2S; sulphuric acid, H_2SO_4;ammonium salts are used for various process operations. Apart from the chemical effluents, 385-545 kg of solid waste is generated for every tonne of hide processed. Most of the salts unabsorbed by the leather end up in the effluent, thus making it a potent environmental contaminant. Sumathi (2003) has pooled characterisation values obtained by various authors, as reported in Table 8.4.

Table 8.4: Physico-chemical Characteristics of Tannery Effluent and Sludge

Parameter	*Effluent*	*Sludge**
Colour	dark brown	green
Odour	offensive	putrescible
pH	6.0 – 12.0	7.6 – 8.46
Electrical conductivity (dSm^{-1})	11.4 – 55.0	2.3 – 20.8
Total suspended solids (mg L^{-1})	400 - 4000	--
Total dissolved solids (mg L^{-1})	4000 - 50000	--
Biological Oxygen Demand (mg L^{-1})	1000 - 20000	--
Chemical Oxygen Demand (mg L^{-1})	2500 - 30000	--
Organic carbon (per cent)	--	7.40 – 9.56
Sodium (mg L^{-1})	2280 - 35000	9917 – 40698
Chlorides (mg L^{-1})	1805 - 15000	287
Sulphates (mg L^{-1})	500 - 5000	40000
Suphide (mg L^{-1})	59 - 158	--
Oil and grease (mg L^{-1})	1.5 – 4.0	--

Parameter	Effluent	Sludge*
Chromium (mg L^{-1})	7.2 - 1329	812 – 16158
Copper (mg L^{-1})	--	13.0 – 58.0
Zinc (mg L^{-1})	--	38.0 – 218.0

* Values are to be read as mg kg^{-1} instead mg L^{-1}

Problem

These salt rich effluents are generally discharged into nearby barren lands or fields, which ultimately contaminate fresh water ecosystems like rivers, ponds and ground water by leaching or seepage or infiltration.

Effect of Soil Quality

In chrome tanning, 276 chemicals and 14 heavy metals are being used, which ultimately reach the soil either by treated effluent discharge or by sludge disposal. Mahimairaja *et al.* (2000) reported that soils around tanning industries of Vellore district is severely contaminated with sodium (Na), Cr and chloride (Cl), as reported in Table 8.5.

Table 8.5: Soil Contamination Observed in Vellore District of Tamil Nadu

Location	Na (mg L^{-1})	Cl (mg L^{-1})	Cr (mg L^{-1})
Ambur	14216 - 77711	21.3 – 45.87	924 – 16731
Vaniyambadi	2405 - 74398	60.9 - 8175	569 - 79865
Reference soil	1022 - 2697	45.4 – 84.0	5.2 – 8.6

Due to the addition of such salts, mean pH, electrical conductivity (EC) in surface and subsurface soils get altered. Table 8.6 reveals the status of pH and conductivity in North Arcot district.

Table 8.6: Range of pH and Conductivity Values in Tannery Belt

Location	pH*	Conductivity (dSm^{-1})*
Ambur	7.96 – 8.57	0.15 – 12.30
	(8.27)	(3.36)
Vaniambadi	7.68 – 8.87	0.43 – 20.6
	(8.28)	(9.94)
Reference soil	7.96 – 8.23	0.32 – 0.59
	(8.14)	(0.48)

* Figures in parentheses indicate mean value.

Source: Ramasamy *et al.*, 2000.

Chromium Content in Tannery Soils

Eighty per cent of the tanneries are chrome based and therefore Cr contamination is always likely in soils around tanneries. The total Cr concentration in soils varied from 341 – 881 ppm, as observed by Sunitha *et al.* (2015) and presented in Table 8.7.

Table 8.7: Chromium Content in Soil and Water in Tannery Belt of Vellore District

Area	*Soil (mg kg^{-1})*	*Water (mg L^{-1})*
Walajapettai	275 - 881	BDL – 36.7
Ambur	16 - 18	6.1 – 8.3
Arcot	7 - 512	5 – 13.5
Vellore	29 - 479	BDL
Tirupathoor	223 - 1059	BDL
Vaniyampadi	126 - 1067	BDL – 1.8
Gudiyatham	92 - 1646	BDL – 4.5

Among soil types, red soil favours 34 per cent higher oxidation of Cr than black soils due to higher pH and manganese dioxide (MnO_2) content, reports Suganthi (2004). Studies on the speciation of Cr indicated that 85 – 99 per cent of Cr was extractable by nitric acid, while NaOH and EDTA extractable Cr constituted 0.5 – 15 per cent.

Effect on Groundwater

Over the years, the groundwater in the areas where tanneries are located, has become intolerably polluted to make it unfit for drinking, irrigation and for general consumption. It has been established that a single tannery can cause pollution of groundwater around a radius of 7 to 8 km (Bhaskaran, 1977). Deterioration in the quality of groundwater occurs due to infiltration of tannery effluent discharges on soils or nearby by water bodies.

In Dindigul tannery areas also, the impact of the effluents is so stupendous that the water becomes unfit for drinking and irrigation. Mondal *et al.* (1998) reported that the TDS of the groundwater is as high as 17,000 mg L^{-1}. Sodium chloride is the major dominant chemical present in groundwater, which makes it unsuitable for drinking and irrigation. Among the dissolved constituents, Na^{2+}, Ca^{2+}, Mg^{2+}, HCO^{-3} and $SO_2{}^{-4}$ are in excess of the levels recommended for either drinking or irrigation (Table 8.8).

Effect on Plants

Tannery effluent is detrimental to crop growth. The prime accused in this case, sodium, interacts with the insoluble calcium and magnesium in the soil complex rendering them soluble and are thus brought into soil solution and removed from the soil; the sodium from the sodium chloride combines with the clay complex to form soda clay. The resulting sodium clay possesses properties detrimental to plant growth. This phenomenon, referred as 'degree of alkalisation' was more than 50 per cent in areas located near tanneries (Iyer *et al.*, 1952).

Table 8.8: Minimum, Maximum, Mean Values of Ions (mg L^{-1}) in Dindigul Groundwater

	tds	*Ca*	*Mg*	*Na*	*k*	*hco$_3$*	*ci*	*sO$_4$*	*NO$_3$*
Minimum	349	38	1	26	1	31	25	13	1
Maximum*	17000	1741	936	4850	215	756	10390	961	252
Mean	2496	288	145	348	21	377	1079	185	35
Standard deviation	2507	307	163	545	34	140	1560	161	44

Source: Mondal *et al.*, 1998.

The effects on crops are seemingly visible in the germination stage itself. Saxena *et al.* (1986) in their studies in Agra observed reduction in germination of pulses *viz.*, pea (33.6 per cent), black gram (15.4 per cent) and green gram (8.2 per cent) and subsequently seedling growth. In rice varieties ADT 37, ADT 39 and red gram (LR 47), the germination and seedling growth was retarded even at 50 per cent dilution, indicating severity of the effluent on plants. Similar effect was pronounced in sunflower also.

However, proper dilution of properly treated tannery effluent was reported to alleviate the impact on plant growth. Karunyal *et al.* (1994) reported improved performance of *Solanum lycopersicum* (tomato), Vigna*sp.*, and *Gossypium hirsutum* (cotton) upon 25 per cent effluent concentration. Similarly, Lentil, *Lens culinaris*; flower crops *Jasminum sambac* and *Neerium oleander*; trees like *Pongamia pinnata, Ficus relegiosa, Terminalia arjuna, Azadirachta indica, Syzigium cumina* and *Acacia nilotica and Casuarina equisetifolia* (Anandhkumar, 1998) were better performers at low effluent concentrations.

Affected Area

An exhaustive study carried out by Tamil Nadu Agricultural University across Palar river basin stated that more than 50,000 ha of agricultural land were polluted and that the yield loss was to the tune of 75 per cent in paddy, 52 per cent in coconut and 48 per cent in sugarcane. The income loss incurred by the farmers was observed to be between 15 and 80 per cent.

The extent of groundwater contamination, as measured by salt load (EC > 3 dSm^{-1}) in Vellore district was more than 45 per cent in Arcot taluk; 46 per cent in Katpadi taluk (Rajannan *et al.*, 1999) and 40 per cent in Walajapettai taluk (Thangavel *et al.*, 2003).

Technological Interventions

Tannery pollution issues can be addressed by a combination of several fool proof technologies to minimise its environmental damage. A combination of aerobic and anaerobic treatment of effluent, soil amendments, microbial remediation, phyto-remediation using macrophytes and growth of tolerant crops/trees are some of the suggested measures to alleviate the problem. Tamil Nadu Agricultural University has conducted detailed research on tannery pollution and suggested the following remedial measures:

- ✰ Application of composted coconut coirpith @ 10 t ha^{-1} in the tannery polluted area was found to be the best soil amendment.
- ✰ Liming the soil above pH 7, renders the Cr less mobile and less bio-available.
- ✰ Suitable crop varieties recommended for tannery affected soils are :

Cereals	: Rice (TRY 1, CO 43, Paiyur 1, ASD 16)
Millets	: Ragi (CO 12, CO 13)
Sugarcane	: Co G (SC) 5
Oilseeds	: Sunflower (CO 4, Morden) and Mustard
Cash crops	: Sugarcane (COG 94076, COG 88123, COC 771)
Vegetables	: Brinjal, Bhendi, Chillies, Tomato (PKM 1)
Flower crops	: Jasmine, Neerium, Tuberose
Trees	: *Eucalyptus, Casuarinas* and *Acacia*

Some other possible options, revealed through various research programmes are:

- ✰ Reed bed treatment system using *Arundo* sp. for Cr removal and BOD, COD reduction

Bacterial cultures *Pseudomonas sp.* and *Bacillus* sp; fungal cultures *Aspergillus niger* and *A. flavus* could reduce Cr toxicity in the tannery effluent when supplemented with 1 per cent glucose and sucrose as carbon source.

II. Paper and Pulp Industrial Pollution Management

The pulp and paper industry plays an integral role in the global economy. India is the 15th largest paper producer in the world. Today, there are 759 pulp and paper mills with an installed capacity of 12.7 million tonnes producing 10.11 million tonnes of paper and paperboards which is 2.52 per cent of the total world production of 402 million tonnes per annum. In Tamil Nadu, there are about 31 paper units with an installed capacity of 0.64 million tonnes producing fine duplex paper boards per annum. The pulp and paper industry is ranked third in the world after primary metals and chemical industries in terms of fresh water use consuming around 250 to 300 m^3 freshwater per tonne of paper produced. The corresponding wastewater generation is also much higher ranging from 75 to 250 m^3 per tonne of paper produced. The characteristics of pulp and paper mill industrial effluents are given in Table 8.9.

Problems

The gradual decline in crop productivity was observed over the years. Majority of the farmers have changed their farming practice of cultivating annual crops (sunflower, tapioca, sugarcane, sorghum) to perennial coconut based intercropping either with fodder grass or annual crops like sugarcane, tapioca, fodder sorghum *etc.* In general, the value of agricultural land was low in industrial areas than the non-industrial areas.

Table 8.9: Characteristics of the Pulp and Paper Mill Industrial Effluent

Sl.No.	Parameters	Treated effluent		
		TNPL	ITC	Seshasayee
1	Colour	Light brown	Colourless	150
2	pH	7.30	7.34	7.1- 7.6
3	EC (dS m^{-1})	1.45	1.75	0.9- 1.3
4	TDS (mg L^{-1})	760	1340	680-710
5	DO (mg L^{-1})	3.42	4.0	--
6	BOD (mg L^{-1})	8.20	18.0	10-14
7	COD (mg L^{-1})	175.0	172.0	--
8	Ca (mg L^{-1})	310.4	265	196-216
9	Mg (mg L^{-1})	57.60	172	90-146

Effect of Pulp and Paper Mill Effluent Irrigation on Groundwater Quality

The ground waters within the effluent irrigated areas had high EC, total hardness, carbonates, bicarbonates, chlorides, sulphates, sodium, calcium, magnesium, potassium, per cent sodium, SAR and RSC than that of cauvery river water. The EC and SAR values increased 2 to 3 times within four years of continuous effluent irrigation.

TNAU Technological Intervention

Different crops, grasses and forest tree species which are directly or indirectly involved in degrading organic and inorganic pollutants (Na, Cl, SO_4 *etc.*) associated with mitigating paper mill effluent polluted soil habitats along with appropriate management strategies for effective restoration and rehabilitation are developed.

TNAU Constructed Wetland Technology

TNAU constructed wetland technology is recommended for treating the pulp and paper mill effluent using species *viz., Typha latifolia, Pharagmitis australis* and *Cyperus pangorei* with plant density of 2.5 lakhs shoots ha^{-1} (25 shoots m^{-2}). Around 1 ha of wetland area is required to treat 1000 m^3 of wastewater per day with a retention time of 2-3 days.

Application of Effluent on Yield and Crop Growth

Effluent water emanate from pulp and paper mill contain good amounts of nutrients and it can be used as a source of irrigation for crops like cereals, millets, oil seeds, pulses and fodder grasses (Ponniah and Ramasamy, 1997). Oil content of groundnut crop was increased due to treated TNPL effluent irrigation than well water. The crude protein content of groundnut was increased due to TNPL effluent

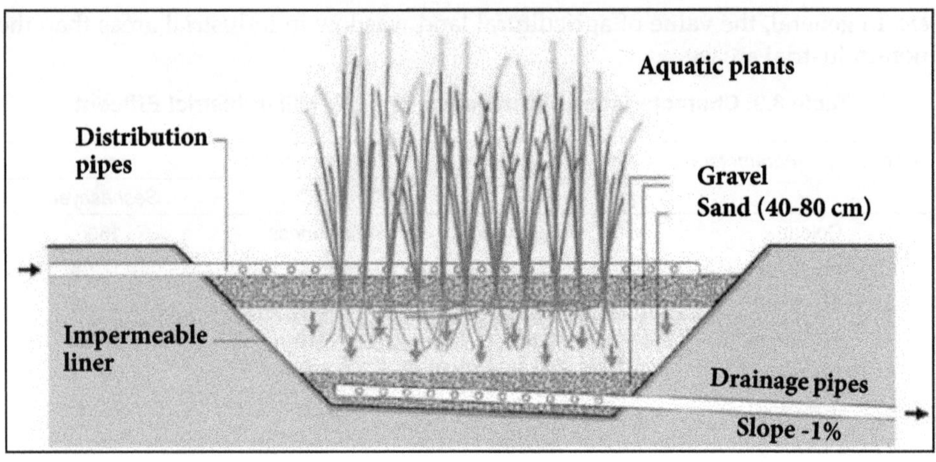

irrigation. The yield of sugarcane increased under continuous treated paper mill effluent irrigated fields to a tune of about 5.13, 3.53, 3.27 and 2.00 per cent over the well water irrigation for the varieties CoC 92061, CoC 671, CoC 6304, respectively. Whereas the cane yield increase was 8.7, 7.4, 7.7, 8.6 and 6.9 for the varieties CoSi 95071, CoC 92102, Co 86032, CoSi 95071 and CoSi 86071, respectively (Udayasoorian *et al.*, 2004).

Soil Amendment

The soil amendments had favourable effects on the P and K uptake of rice in acid soil. The use of 50 per cent diluted classified effluents was found superior to raw effluent irrigation more particularly in neutral soil. The adverse effects of effluent from paper factory could be alleviated by resorting to the application of N, P and K along with organic and inorganic amendments such as pressmud, FYM and gypsum (Pushpavalli, 1990).

Mahar *et al.* (2008) reported that application of pressmud @ 25 t ha^{-1} on sugarcane recorded higher number of tillers and sugar recovery after inorganic fertilizer application @ 225- 112-168 kg ha^{-1} of NPK. Enriched press mud @ 10 t ha^{-1} also increased the number of millable canes in sugarcane (Rakkiyappan *et al.*, 1999). The increase in brix, pol per cent, purity and CCS of sugarcane juice were 1.03, 0.97, 0.26 and 0.43 units, respectively under effluent irrigation than under well water irrigation. The application of FYM, biocompost, vermicompost and fly ash under treated effluent irrigation had favorable effects on growth characters and yield attributes *viz.*, length, girth and weight of beetroot. The experimental results revealed that yield of beetroot under treated effluent irrigation were significantly higher than well water irrigation. The treatment combination of 100 per cent NPK + vermicompost @ 3.5 t ha^{-1} + fly ash @ 5 t ha^{-1} performed better.

III. Textile and Dyeing Industrial Pollution Management

Textile and dyeing industries have great economic significance by virtue of its contribution to national economy and economic generation. There are about 2050 textile and dyeing industrial hubs in India. Tirupur is one such hub located in Tamil Nadu, having about 750 dyeing and bleaching units.

Textile industry is a very diverse sector covering various processes like bleaching, dyeing, printing and stiffening. These industries generate large quantities of wastewater and solid wastes, which are polluting land and water severely in and around Tirupur. About 75 - 100 million litres of effluent is being disposed onto soil and water systems, affecting more than 12000 ha of agricultural lands in Tirupur District (Table 8.10). This effluent is discharged onto Noyyal river which has affected several thousand hectares of agricultural lands in Coimbatore, Erode and Karur Districts also. The entire Noyyal river basin was found to be contaminated with high salt load, making soils in the area unfit for agriculture.

Its biggest impact on the environment is related to primary water consumption and waste water discharge (115–175 kg of COD/ton of finished textile), wide range of organic chemicals, low biodegradable compounds, colour and salinity.

Table 8.10: Land Area Affected Due to Dye and Textile Effluents in Tirupur District

Moderately Affected EC 1.5 - 3.00 (dS/m)	*Severely Affected EC 3.00 - 5.25 (dS/m)*	*Very Severely Sffected EC 5.25 - 7.5 (dS/m)*	*Extremely Affected EC > 7.5 (dS/m)*	*Total Area (ha)*
3472.23	1742.26	886.59	740.61	6841.79

Table 8.11: Key Characteristics of Textile Industry Effluent

pH	6 – 9
TSS (mg L^{-1})	200 – 1875
TDS (mg L^{-1})	2469 – 7295
BOD (mg L^{-1})	125 - 653
COD (mg L^{-1})	115 - 705

Source: Varsha *et al.*, 2013.

Sesuvium portulacastrum, a coastal plant is found to tolerate abiotic constraints such as salinity and drought. It was observed to grow at severe salinity of 1000mM NaCl, indicting its prospects in scavenging sodium from textile contaminated soils.

IV. Impact of Distillery Spentwash on Soil and Water Quality

India is a major producer of sugar in the world, and sugar industry offers employment potential and contributes substantially to economic development. There are 579 sugar mills and 356 distilleries in India. Apart from the sugar and alcohol, these factories generate many by-products and waste materials. These industries generate about 80 million tons of molasses as valuable by products (Economic times, 2015).

Currently, about 40 million m^3 of spent wash is generated annually from distilleries alone in India. The raw spentwash is acidic in pH (3.94 - 4.30), high EC

(30 - 45 dSm^{-1}), Nitrogen (1660 - 4200 mg L^{-1}), phosphorus (225 - 3038 mg L^{-1} and potassium (9600 - 17475 mg L^{-1}). It also contains considerable quantities of calcium, magnesium, sulphate, chloride, zinc, copper, iron and manganese. Thus, it can effectively be used as a source of plant nutrients and as soil amendment. The high concentration of Ca (2050 – 7000 mg L^{-1}) could be exploited for reclaiming the sodic soils similar to that of gypsum effect. Table 8.12 depicts the nutrient potential of spentwash.

The spentwash generally does not contain any toxic metals, but is characterized by a high biological oxygen demand (>40000 mg L^{-1}) and chemical oxygen demand (>80000 mg L^{-1}).

Table 8.12: Nutrient Potential of Distillery Spentwash in Major States of India

Sl.No.	*State*	*Distillery Units*	*Capacity (MLtr/yr)*	*Total DSW Generated (M Ltr/yr)*	*Total N (tones/yr)*	*Total P (tones/yr)*	*Total K (tones/yr)*
1	Andhra Pradesh	29	193.559	2323	2787	697	27872
2	Gujarat	15	171.05	2053	2463	616	24631
3	Karnataka	40	374.514	4494	5393	1348	53930
4	Madhya Pradesh	18	308.051	3697	4436	1109	44359
5	Maharashtra	81	849.222	10190	12228	3057	122285
6	Punjab	13	256.847	3082	3699	925	36986
7	Tamil Nadu	26	416.425	4997	5997	1500	59965
8	Uttar Pradesh	53	1137.482	13650	16380	4095	163797
Overall status in India (including all states)		**356**	**4229.694**	**50756**	**60907**	**15227**	**609073**

Environmental Impact of Spentwash

The contamination of surface and groundwater, destruction of aquatic life, salt accumulation in soil, lowering of pH value of the stream, increase in organic load, depletion of dissolved oxygen, discolouration and bad smell are some of the major pollution problems associated with the disposal of distillery spentwash. During rains and/or irrigation water, the salts and nutrients in spentwash polluted soils leach down through the soil profile tand contaminate the groundwater. There are three major effects of these salts on plant growth *viz.*, a. direct toxicities of salts; b) ionic imbalances in the soil and plant; c) reduction in the availability of water by lowering the osmotic potential. This has been termed physiological drought because plants are affected by lack of water even though the water content of the soil is apparently adequate for crop needs.

Another major problem in distillery spentwash is colour. The distillery spentwash has brown to very dark brown colour, which is of organic nature. Characterisation of the colouring principle of spentwash by Samuel Raj and Ramasamy (1997) showed the presence of melanoidin compound with high similarity to synthetic melanoidin. The colour poses a serious threat to the environment as the water bodies receiving coloured waste might got coloured affecting the penetration of light which damages the ecosystem productivity.

Management Strategies for Distillery Spentwash

The hazards associated with spentwash can be minimized using appropriate management technologies based on the scientific principles. The department of Environmental sciences through many collaborative projects with different sugar and distillery industries located all over Tamil Nadu and conducted many lab and field studies for optimizing the dosage of spentwash to different crops and to improve the soil fertility. In addition to this, technologies were also developed to monitor and assess the impact of spentwash application on water quality.

Technological Interventions

Reclamation of Sodic Soils

Sodicity is a problem in soils posing a serious threat to the permanence of irrigation agriculture. Amendments generally used to reclaim sodic soils are gypsum, phosphogypsum, iron pyrites and elemental sulphur. All these are inorganic in nature. Some of the organic amendments to reclaim the sodic soils are press-mud, farmyard manure (FYM), coir dust and green manures. The direct discharge of untreated distillery effluent (spentwash) to reclaim and improve the productivity of the sodic soils is advocated. The acidic nature and high SO_4 in spentwash make it an amendment for reclamation of sodic soil similar to gypsum.

Hence, only one time application of 3.75 to 5.00 lakhs litres of untreated distillery effluent (spentwash) per hectare of sodic soils in summer months is recommended (CPG, 2012). Natural oxidation can be induced for a period of six weeks with two intermittent dry ploughing at a particular interval. Then, after 45 – 60th day of application, soil is to be irrigated with fresh water and drained. This treatment reduces the pH and exchangeable sodium percentage to normal level and increases

the productivity of the sodic soils. After this reclamation practice, rice crop can be raised in the effluent applied field adopting the conventional cultivation technique. Application of this effluent again to the next crop/season or year after year and also to the land nearby drinking water sources is not advocated.

Distillery Spentwash as Liquid Fertilizer to Crops

As spentwash contains almost all plant nutrients, it could be used as liquid manure. Fertilizer application through irrigation water is being recommended to improve the fertilizer use efficiency. Since the effluent has higher dissolved salts, 50 times diluted effluent can be irrigated to sugarcane, banana, ragi, sunflower, grasses, cotton and soybean. It can also be used as one time application to fallow land at the rate 20,000 to 40, 000 litres per hectare. It should be allowed for complete drying over a period of 20 to 30 days. The effluent applied field is to be thoroughly ploughed two times for the natural oxidation and mineralization of organic matter. After that, crops can be raised in the effluent applied field adopting the conventional methods. Application of this effluent again to the next crop/season or year after year and also to the land nearby drinking water sources is not advocated. Spentwash at 5 and 10 per cent concentration can be recommended for Seed hardening to improve the germination, root and shoot length in the crops *viz.*, maize, sunflower, red gram, cotton, white sorghum, cumbu and groundnut.

V. Sewage Irrigation and Soil Fertility

Nearly, about 70–80 per cent of water utilized for domestic purposes is discharged as wastewater or sewage. The CPCB has estimated that India's major cities and municipalities generate 38254 lakh litres as sewage every day. While majority of this water is used for agriculture, remaining is released either as treated or untreated. This cause pollution of land and water and also contaminate food produced in such lands. The characteristics of sewage are furnished in Table 8.13.

Effect of Sewage Irrigation on Soil

Application of sewage for irrigation increases bulk density, structure, water holding capacity and infiltration capacity. It also increases the pH status and salt content of soils. Continuous irrigation of sewage to soils increases soil organic carbon, which plays a major role in soil fertility.

Rice cultivation over 2100 ha along 10 km stretch on both sides of River Moosi in Hyderabad is done. Wheat cultivation is also being practiced in Ahmedabad and Kanpur. In Delhi, Bhendi, Cucumber, Brinjal, coriander, Spinach, mustard, cauliflower, cabbage are cultivated in an area of 1700 ha using sewage. Also, sewage is used for jasmine cultivation and garden plants in Kanpur and Hyderabad. In Moori river basin, fodder crops is being grown in an area of 10000 hectares.In rural Hoobly and Dharwad areas of Karnataka, farmers use sewage water for raising sapota, guava, cocnut, mango, arecanut, teak, neem, banana, curry leaf, pomegranate, lime and mulberry.

Table 8.13: Characteristics of Treated Sewage

Sl.No.	Properties	Values
1	pH	7.3 ± 0.09
2	EC (dS m^{-1})	0.78 ± 0.34
3	Calcium (mg L^{-1})	4.76 ± 0.46
4	Magnesium (mg L-1)	1.69 ± 0.032
5	Sodium (mg L^{-1})	1.42 ± 0.022
6	Potassium (mg L^{-1})	0.31 ± 0.07
7	Iron (mg L^{-1})	83.0 ± 6.24
8	Manganese (mg L^{-1})	22.0 ± 0.42
9	Zinc (mg L^{-1})	2.0 ± 0.054
10	Copper (mg L^{-1})	2.4 ± 0.010
11	Lead (mg L^{-1})	7.5 ± 0.32
12	Nickel (mg L^{-1})	4.1 ± 0.014
13	Cadmium (mg L^{-1})	2.1 ± 0.045

Source: Singh, 2012.

Because, improperly treated sewage contaminates land, water and ground water thereby facilitating contaminants in foods that are being grown in sewage irrigated soils. Heavy metals like chromium, cadmium, mercury, copper and antibiotic residues are present in high concentrations in sewage. Since, heavy metal are not fully removed using conventional treatment plants, sewage has to be used for irrigation only after proper treatment and removing these pollutants. Hence, the farmers should ensure that whether the water they are using is properly treated before irrigation. Farmers are advised to use such waters for flower crops, fodder crops, trees, gardens, lawns, orchards and afforestation programmes rather than being used for growing edible crops. If properly done, sewage irrigation not only reduces consumption of fresh water for agriculture, but also improves soil nutrient status and higher productivity.

References

Anandhkumar,S.P. 1998. Studies of treated tannery effluent on flower crops and its impact on soil and water quality. M.Sc. thesis submitted to Tamil Nadu Agricutural University, Coimbatore, India.

Bhaskaran, 1977. Treatment and Disposal of Tannery Effluents. CLRI, Chennai

Bosnic, M., J. Buljan, R. P. Daniels and S. Rajamani. 2003. Pollutants in tannery effluent, International scenario on environmental regulations and compliance. Report submitted to UNIDO, p. 27.

CGWB, 2011. Groundwater Year Book, India, Central Groundwater Board, Ministry of Water Resources, Govt. of India.

Crop Production Guide, 2012. Department of Agriculture, Chepauk, Chennai, pp. 33-147.

Iyer, C.R., R.Rajagopalan and S.C.Pillai. 1952. Soil conditions as affected by tannery waste waters. Journal of the Indian Institute of Science, 34: 163 – 178 - http: // journal.library.iisc.ernet.in/index.php/iisc/article/view/13253

Iyer,V.G. and N.E. Mastorakis. 2006. Assessment of pollution load from unsafe leather tanneries in India. WSEAS Transactions on Environment and Development, 2(3): 207-215.

Kaul, S.N., T, Nandy, A.Gautam, L. Szpyrkowicz and D.R.Khanna. 2005. Wastewater Management with Special Reference to Tanneries. Discovery publishing house, 467 pp

Mahar, G.M., U.A. Burino, F.C. Oad and S.A. Shaikh. 2008. Cane yield and sugar recovery of sugarcane variety Larkama – 2001 under different fertilizer sources. Asian J. Plant Sci.,

Mahimairaja, S, S.Sakthivel, J.Divakaran, R.Naidu and K.Ramasamy. 2000. Extent and severity of contamination around tanning industries in Vellore distict. In: Proceedings of the International Workshop on 'Towards Better Management of Soils Contaminated with Tannery Waste' (Naidu, R., I.R.Willeet, S.Mahimairaja, R.Kookana and K.Ramasamy eds), held at Tamil Nadu Agricultural University, Coimbatore, 31 January - 4 February, pp: 75 – 82.

Mondal, N.C., V.K.Saxena and V.S.Singh. 1998. Impact of pollution due to tanneries on groundwater regime. Curr.Sci., 88 (12): 1988-1994.

Ponniah, C and K. Ramasamy. 1999. Impact of paper mill effluent irrigation in fodder grass soil ecosystem. **In:** Proc. the workshop on bioremediation of polluted habits. Tamil Nadu Agric. Univ., Coimbatore, pp. 62-67.

Pushpavalli, R. 1990. Studies on the characterization of pulp and paper mill effluent and its effect on soil profile characteristics and on germination, yield and juice quality of sugarcane. M.Sc. dissertation, Tamil Nadu Agric. Univ., Coimbatore, India.

Rakkiyappan, P., M.V. Navaneethakrishnan and B.K. Tripathi. 1999. Effect of enriched PM (bioearth) and biofertilizers on scans. **In:** Int. Conference on Contamination in the soil environment in the Australia – Pacific region, Dec. 12-17, New Delhi, pp.41-42.

Ramasamy,K., S.Mahimairaja and R.Naidu. 2000. Remediation of soils contaminated with chromium due to tannery wastes disposal. In: Remediation Engineering of Contaminated Soils (Wise, D.L. ed.), CRC press, Sydney, pp: 583 – 614.

Samuel Raj, H and P. Ramasamy, 1997. Characterization of colouring principle – Environment issues related to agriculture, textile and leather. **In:** Proceedings of the 6th national symposium on environmental issues related to agriculture, textile and leather, TNAU, Coimbatore, pp. 286-290.

Sankar, U. 2001. Economic Analysis of Environmental Problem in Tanneries and Textile Bleaching and Drying Units and Suggestions for Policy Action. Allied publishers Ltd., New Delhi, P. 298.

Saxena, R. M., P.F. Kewal, R.S. Yadav and A.K. Bhatnagar. 1986. Impact of tannery effluents on some pulse crops. Indian J Environ. Hlth., 28(4): 345-348.

Sengupta, A.K. 2008. WHO guidelines for the safe use wastewater, excreta and grey water.In: Proc. of National Workshop on Sustainable sanitation, New Delhi, 19-20, May, 2008.

Singh, R.A.2012. Impact of sewage irrigation and industrial effluent on soil, plant and health. Industrial wastes (Show, KY ed.), Intech, pp. 49-72.

Suganthi, C. 2004. Behaviour of chromium in CETP tannery sludge applied soils and effect on groundnut crop. Thesis submitted to Tamil Nadu Agricultural University for the award of Doctor of Philosophy, p.89.

Sumathi, K.M.S. 2003. Reed bed system for treating tannery effluent to minimize environmental impact of chromium. Thesis submitted for the award of Doctor of Philosophy to Tamil Nadu Agricultural University, p.163.

Sunitha, R., G.Purushothaman, A.Bharani and S.Mahimairaja. 2015. Chromium contamination in soil and groundwater due to tannery wastes disposals at Vellore district of Tamil Nadu. International Journal of Environmental Sciences, 6 (1): 114-124.

Thangavel, P., Rajannan, G., Ramesh., P.T. and K.Ramasamy. 2003. Assessment of quality of well water and soils nearing tannery areas in Walaja taluk. *Journal of Ecotoxicology and Environmental Monitoring* 13: 17 – 20.

Udayasoorian, C., P. C. Prabu and K. Mini. 2004. Influence of composted bagasse pith and treated paper mill effluent irrigation on groundnut. Madras Agric. J., 91: 126-129.

Varsha, G., Sudesh and S.Sena. 2013. Physico-chemical analysis of textile effluent of dye and printing clusters of Bagru region, Jaipur, India. India J. Environ. Res. Develop., 8(1): 11-15.

Chapter 9

Yield and Quality Parameters Up Scaling through Crop Boosters

R. Sivakumar and P. Jeyakumar

Department of Crop Physiology,
Tamil Nadu Agricultural University,
Coimbatore – 641 003, Tamil Nadu

The Department of Crop Physiology, TNAU, Coimbatore has the credit of production and distribution of TNAU Coconut Tonic to the coconut growers in Tamil Nadu and adjoining states of Kerala, Karnataka and Andhra Pradesh for the last 12 years. Following the success of TNAU Coconut Tonic, the Department of Crop Physiology has developed few more crop boosters to improve yield in pulses, sugarcane, cotton, maize and groundnut. These boosters increase the growth, yield and quality of crop. The Crop Boosters available in the Department of Crop Physiology include TNAU Coconut Tonic, TNAU Pulse Wonder, TNAU Groundnut Rich, TNAU Cotton Plus, TNAU Maize Maxim and TNAU Sugarcane Booster.

TNAU Coconut Tonic

Coconut is known for its great versatility as seen in the many uses of its different parts and found throughout the tropics and subtropics. Coconuts are part of the daily diets of many people. Coconuts are different from any other fruits because they contain a large quantity of "water" and when immature they are known as tender-nuts or jelly-nuts and may be harvested for drinking. When mature, it can be used as seed nuts or processed to give oil from the kernel, charcoal from the hard shell and coir from the fibrous husk.

Of the primary nutrients, potassium has been found to be the most important in coconut cultivation, followed by nitrogen. There is a general response to the application of K and N; while response to phosphorus is seen only in certain restricted and localized condition. Among the secondary nutrients, magnesium and chlorine have beneficial effects, followed by calcium, sulphur and sodium. Among micro-nutrients, zinc, boron, molybdenum and manganese are required under certain restricted conditions.

Importance of Nutrients in Coconut

Nitrogen is of great importance for the rapid development and growth of the trees. Nitrogen promotes the developments of the vegetative parts of the plant especially the leaves and shoots as well as to increase the number of leaves. Phosphorus is found especially in leaves and seeds and also in parts of the plants where vigorous cell division is taking place. It plays an important role in root growth and increased yields. Potassium is particularly necessary for the formation of sugar, fat, and fibrous material, the coconut palm may be expected to have a high requirement of potassium. Magnesium improved production of more female flowers, high setting percentage and more number of nuts per bunch. It also plays an important role in photosynthesis and greenness of leaves. Sulphur increases the oil content. Boron is more important for fertilization and formation of nuts.

Deficiency Symptoms

Nitrogen deficiency begins as a uniform light green discoloration/yellowing (uniform chlorosis) of the oldest leaves. Yellowing starts from tip to base of the lower leaves and will proceed up. Phosphorus deficiency shows purple coloration in leaves (In severe cases, the leaves turn yellow before drying prematurely), leaves stay upright and premature leaf shedding.

Boron deficiency symptoms always occur on newly emerging leaves, and remain visible on these leaves as they mature and are replaced by younger leaves. One of the earliest symptoms of B deficiency on coconut palm is leaf wrinkling and manifested as sharply bent leaflet tips, commonly called "hook leaf". These sharp leaflet hooks are quire rigid and cannot be straightened out without tearing the leaflets. Leaves have a serrated zigzag appearance. Boron deficiency also occurs in inflorescence and nuts. The inflorescence and nuts are become necrotic and shedding of buttons. Along with nutrients, some hormone like auxin is essential for growth and button development.

Common nutritional disorders in coconut are Pencil point disorder, Button shedding, Small nuts, Nut splitting and Barren nut. These disorders can be corrected by using TNAU Coconut Tonic.

TNAU Coconut Tonic

- ☆ A liquid booster with nutrients and growth regulators essential for Coconut.
- ☆ Application of coconut tonic to the tree through root feeding technique.

- ☆ Root feeding of the tonic @ 200 ml/tree - twice per year at six months interval.
- ☆ Just two to three feet away from the trunk, the soil has to be stirred with a fork or a pick axe, and a white coloured feeding root of pencil thickness is located. A slanting cut is given at the tip with a sharp knife or a blade, and the root is inserted into the polythene bag containing the tonic. The bag and the root are tied with a thread at the top of the bag.

Benefits

- ☆ Increases chlorophyll content and greenness of leaves
- ☆ Improves photosynthetic efficiency of leaves
- ☆ Decreases button shedding
- ☆ Increases number and size of nuts
- ☆ Increases nut yield up to 20 per cent
- ☆ Increases longevity and vigour of the palm
- ☆ Imparts resistance to pests, diseases and environmental stresses

TNAU Pulse Wonder

Pulses are rich in proteins and found to be main source of protein to vegetarian people of India. It is second important constituent of Indian diet after cereals. Out of 17 essential elements, pulses specially need adequate amount of P, Ca, Mg, S and Mo. Phosphorus is required for proper root growth and growth of rhizobia. Calcium and magnesium are required to stimulate growth and to increase the size of the nodules, pod formation and grain setting. Sulphur is required for nodulation and protein synthesis. Molybdenum for nitrogen fixation and assimilation and boron for reproduction are the requirement. Flower shedding, premature leaf shedding and poor yield are common nutrient disorders in pulse crops. TNAU Pulse Wonder is a booster with nutrients and growth regulators essential for Pulses.

Application

- ☆ Dose: 2 kg/acre
- ☆ Spray volume: 200 litres
- ☆ Stage of spray: Peak flowering
- ☆ Add adequate quantity of wetting agent

Benefits

- ☆ Decreases flower shedding
- ☆ Increases yield up to 20 per cent
- ☆ Increases drought tolerance

TNAU Groundnut Rich

Groundnut is one of the most important cash crops of our country. It is a low-priced commodity but a valuable source of all the nutrients. Groundnut is the sixth most important oilseed crop in the world. It contains 48-50 per cent of oil and 26-28 per cent of protein, and is a rich source of dietary fiber, minerals, and vitamins. Groundnut is called as the 'king' of oilseeds. It is one of the most important food and cash crops of our country. Groundnut is also called as wonder nut and poor men's cashew nut. Among the nutrients, Fe, B, Ca, P and K are mainly required for its growth, development and yield. Pod filling is a major problem especially in the bold seed varieties. To improve pod filling spraying of nutrient solution is unavoidable.

The major nutrient deficiency disorders are Hollow heart, Heart rot and Chlorosis in younger leaves. TNAU Groundnut Rich is a booster with nutrients and growth regulators essential for Groundnut.

Application

- ✰ Dose: 2 kg/acre/spray
- ✰ Spray volume: 200 litres/spray
- ✰ Stages of spray: Peak flowering and pod development
- ✰ Add adequate quantity of wetting agent

Benefits

- ✰ More flower retention
- ✰ Improves pod filling
- ✰ Increases pod yield up to 15 per cent
- ✰ Improves drought tolerance

TNAU Cotton Plus

In many countries cotton is one of the most important fibre producing plants. Cotton crop not only provides fibre for the textile industry, but also plays a role in the feed and oil industries with its seed, rich in oil (18 – 24 per cent) and protein (20–40 per cent). Nutrient management in cotton is complex due to the simultaneous production of vegetative and reproductive structures during the active growth phase. The nutrient demand by the fruiting parts is very high. High nutrient demand at this stage results in reduction of root growth due to less partitioning of assimilates to the root and ultimately reduced capacity to absorb nutrients. An excess of nutrient applied, especially N before the crop attains the grand growth period could revert the crop to putting up more of vegetative growth. A deficiency could result in hastening maturity. Cotton being deep rooted removes large quantities of nutrients from the soil profile. For every 100 kilogram of seed cotton produced the crop depletes the soil by 6-7 kg N, 1.9-2.5 kg P, 6-8 kg K and 1.2-2.0 kg S.

Among the nutrients, N, P, K, Mg, B, S and Fe requirement is high for cotton growth and yield. Common disorders produced by the cotton crop due to nutritional deficiency are Reddening, Square and boll shedding, small fruits and poor bursting of bolls. TNAU Cotton Plus is a booster with nutrients and growth regulators for Cotton.

Application

- ✰ Dose: 2.5 kg/acre/spray
- ✰ Spray volume: 200 litres/spray
- ✰ Stages of spray: Flowering and boll formation
- ✰ Add adequate quantity of wetting agent

Benefits

- ✰ Reduces flower and square shedding
- ✰ Improves boll bursting
- ✰ Increases seed cotton yield up to 18 per cent
- ✰ Increases drought tolerance

TNAU Maize Maxim

Maize has becoming very popular cereal crop in India because of the increasing market price and high production potential of hybrid varieties in both irrigated as well as rainfed conditions. More ever in irrigated areas farmers produce the income equal to the cash crops such as sugarcane, onion, cotton, *etc.* in comparatively short time period of 120-130 days by cultivating hybrid maize varieties. Hence, the trend of replacing some cash crops with maize in intensive cultivation is observed in present condition. For increasing the profitability of maize in only economic view, farmers are cultivating the crop intensively with the huge use of chemical fertilizers, pesticides, weedicides, *etc.* Maize crop has better yield response to chemical or inorganic fertilizers. Hence, heavy doses of these fertilizers are applied to maize. Though these practices are helps to increase the temporary increase the production of crop; deterioration of natural resources (*viz.* land, water and air) is also the another side of such high input intensive cultivation. Over reliance on use of chemical fertilizers has been associated with declines in soil physical and chemical properties and crop yield and significant land problems, such as soil degradation due to over exploitation of land and soil pollution caused by high application rates of fertilizers.

Hence, foliar application of nutrients is unavoidable in maize crop. TNAU Maize Maxim is a booster with nutrients and growth regulators essential for Maize. The common nutrient disorders are White bud (Zn deficiency), Ill filled grains in cob tip (N and K deficiency and Irregular grain filling (B deficiency). These disorders can be corrected by using TNAU Maize Maxim.

Application

- ✰ Dose: 3 kg/acre/spray
- ✰ Spray volume: 200 litres/spray

- ☆ Stages of spray: Tassel initiation and grain filling stages
- ☆ Add adequate quantity of wetting agent

Benefits

- ☆ Improves grain filling
- ☆ Increases grain yield up to 20 per cent
- ☆ Improves drought tolerance

TNAU Sugarcane Booster

Sugarcane being a long duration, exhaustive crop removes considerably higher amount of plant nutrients from the soil. On an average, sugarcane crop, yielding 100 tonnes, removes 208, 53 and 280 kg per hectare of N, P_2O_5 and K_2O, respectively from the soil. Hence, it is essential to replenish the depleted soil with plant nutrients at desired levels to restore and sustain the fertility/productivity of soils through integrated nutrient management system.

Among various reputes of sugarcane production, nutrients contribute maximum to the increase in yield. During vegetation the sugarcane consumes many nutrients. There is no doubt that sugarcane crop needs nutrients. The most active uptake of nutrients is observed during the early stage of the sugarcane plant, during tillering (from the third to the sixth months after planting). Among the nutrients, nitrogen required for vegetative growth (tillering, foliage formation, stalk formation and growth) and root growth. Vegetative growth in sugarcane is directly related to yield. Phosphorus is essential for root growth and necessary for adequate tillering. Potassium is important for sugar synthesis and translocation to the storage organs and develops resistance to sugarcane against pest, disease and lodging. Sulphur increases the juice quality, CCS per cent and cane yield. Zinc is essential for bio synthesis of plant growth regulator auxin. Boron is necessary for cane development, sugar transport, and hormone development.

The common nutrient disorders are Shortened Internodes (Zinc deficiency) and Pahla blight (Manganese deficiency). TNAU Sugarcane Booster is a booster with nutrients and growth regulators essential for Sugarcane.

Application

- ☆ Dose: 1, 1.5 and 2 kg/acre
- ☆ Stages of spray: 45, 60 and 75 days after planting respectively
- ☆ Spray volume: 200 litres/spray
- ☆ Add adequate quantity of wetting agent

Benefits

- ☆ Enhances cane growth and weight
- ☆ Improves internodal length

- ☆ Improves cane yield up to 20 per cent
- ☆ Improves sugar content
- ☆ Increases drought tolerance

Chapter 10

Quality Seed Production: A Basic Critical Input for Sustainable Agriculture

P. Selvaraju[1] and V. Manonmani[2]

[1]Special Officer (Seeds) and [2]Professor
Department of Seed Science and Technology
Seed Centre, Tamil Nadu Agricultural University,
Coimbatore – 641 003, Tamil Nadu

Seed is the basic and most critical input for sustainable and productive agriculture. The response of all other inputs depends on quality of seeds to a large extent. It is estimated that the direct contribution of quality seed alone to the total production is about 15 - 20 per cent depending upon the crop and it can be further raised up to 45 per cent with efficient management of other inputs. In the significant advances that India made in agriculture in the last four decades, the role of the seed sector has been substantial. The expansion of seed industry has occurred in parallel with growth in agricultural productivity. Given the fact that sustained growth to cope with increasing demand would depend more and more on the pace of development and adoption of innovative technologies and the seed would continue to be a vital component for decades to come. The organized seed industry of the country is just forty years old. Yet, its growth has been phenomenal.

The yield gap can be bridged by: (i) Ensuring availability and efficient use of water, fertilizer and plant protection measures, (ii) timely planting of quality seeds and ensuring desired plant population and (iii) development and release of more productive varieties. The modern technological advances have brought about the quality consciousness among the farmers. The good quality seeds are the "Seeds of Green revolution" and the concept of "Seed quality" is a complex one. The efforts of plant breeding will go waste if the quality seed of the improved cultivars/hybrids is not made available to the farmers.

Importance of Seed

Seed is the vital input in crop production because through seed only the investment made on other inputs like pesticide, fertilizer, irrigation and crop maintenance can be realized. The seed required for raising the crop is quite small and its cost is also less compared to other inputs, but the greater income farmer gets depends upon the quality of the small quantity of seed he uses. The seed forms only a small part of the total cultivation expenses, but it will pay its dividend towards yield.

Characters of Quality Seed

- Seed should be physically pure.
- Seed should be genetically pure.
- Seed should be germinable.
- Seed should be highly vigourous.
- Seed should be free from insect and fungal infection
- Seed should have optimum moisture content (low as 6 – 10 per cent for orthodox seed, high as 17 - 30 per cent for recalcitrant seed).
- Seed should have high storability.

Good quality seed is the seed with required genetic and physical purity that is accompanied with physiological soundness and health status.

Physical Quality

It is the cleanliness of seed free from other seeds, debris, inert matter, diseased seed and insect damaged seed. Lack of these quality characters will indirectly influence the field establishment and planting value of seed.

Higher Physical Purity for Certification

- All crops (most): 98 per cent
- Maize, Bhendi: 99 per cent
- Carrot: 95 per cent
- Sesame, Soybean and Jute: 97 per cent
- Groundnut: 96 per cent

These quality characters could be obtained in seed lots by proper cleaning and grading of seed (processing) after collection and before sowing/storage.

Genetic Purity

It is the true to type nature of the seed. *i.e.*, the seedling/plant/tree from the seed should resemble its mother plant in all aspects. This quality character is important for achieving the desired goal of raising the crop either for yield or for resistance or for desired quality factors.

Higher Genetic Purity for Certification

- ✰ Breeder/Nucleus: 100 per cent
- ✰ Foundation seed: 99.0 per cent
- ✰ Certified seed: 98.0 per cent

Seed Production

In seed production, adequate care is given from the purchase of seeds up to harvest adopting proper seed and crop management techniques. The benefits of seed production are:

1. Higher income and
2. Higher quality seed for next sowing.

Maintenance of Genetic Purity during Seed Production

Breeder Seed

Breeder seed is seed or vegetative propagating material which is directly controlled by the originating or in certain cases, the sponsoring breeder or institution and which provides for the initial and recurring increase of foundation seed.

Foundation Seed

Foundation seed is seed stock so handled as to most nearly maintain specific genetic identity and purity. Production is carefully supervised and certified by the seed certification agency. Foundation seed is the source of certified seed class.

Certified Seed

Certified seed is the progeny of foundation or certified seed that is so handled to maintain satisfactory genetic identity and purity and that has been approved and certified by the certifying agency.

Generation System of Seed Multiplication

i. **Three** - Generation model - Breeder seed - Foundation seed - Certified seed

ii. **Four** - Generation model - Breeder seed - Foundation seed (I) Foundation seed (II) - Certified seed

iii. **Five** - Generation model - Breeder seed - Foundation seed (I) - Foundation seed (II) - Certified seed (I) - Certified seed (II).

For most of the often cross pollinated and cross pollinated crops generation model is usually suggested for seed multiplication. For self-pollinated crops, we can go for 4 or 5 generation models, if seed demand is more and seed multiplication is less.

Table 10.1: Generation System of Seed Multiplication and Quality Control (Notified varieties and hybrids)

Agency	*Class of Seed*	*Quality Control System*
Breeder concerned or Sponsoring institution or Breeder himself ←	NUCLEUS SEED (no specified tag) ↓	→ Maintenance breeding
Breeder concerned ←	BREEDER SEED (Golden yellow tag) ↓	→ Breeder seed monitoring team
State Department of Agriculture, National Seeds Corporation, Cooperative agencies, Central and State Seed Corporations, Private Sectors ←	FOUNDATION SEED Stage I and II (White tag) ↓	→ State Seed Certification Agency Field inspection and testing to check minimum required standards of genetic, physical purity and other quality standards
-do-	Certified seed (Azar blue tag) ↓	-do-
	FARMER	

Preceding Crop Requirements

This has been fixed to avoid contamination through volunteer plants and also the soil borne diseases.

Isolation Distance

Distance maintained between the seed crop and sources of contaminants. Isolation distance depends on:

1. Pollination behavior
2. Pollinating agents and
3. Pollen viability.

Seed Certification

The genetic purity in commercial seed production is often maintained through a system of seed certification. The principal objective of seed certification is to maintain and make available to the farmers crop seeds, tubers, or bulbs which are of good

Table 10.2: Minimum Isolation Requirements of Important Crops

	Minimum Isolation Distance (metres)	
Crop	*Foundation Seed*	*Certified Seed*
Paddy, wheat, barley, soybean and groundnut	3	3
Pearl millet	1000	200
Sorghum		
Hybrid	300	200
Open-pollinated variety	200	100
From Johnson grass	400	400
Maize		
Inbred line single cross	400	-
Hybrid maize	-	200
Maize – composites, open pollinated varieties	400	200

seeding value and true to variety. To accomplish these purposes, qualified and well experienced personnel of seed certification agency carry out field inspections at appropriate stages of crop growth. They also make seed inspections to verify that the seed crop/seed lot is of the requisite genetic purity and quality, after harvesting to verify quality, and at the processing plants draw samples for seed testing and sometimes for grow-out tests also. In addition to inspections, seed certification agency also lay down the field and seed standards to which the seed crop and seed lot respectively must conform to get approval as certified seed. The field standards include land requirements, isolation requirements, maximum permissible off-type, shedding tassels (in case of hybrid maize production), *etc.* The genetic purity of seed is thus assured if the certification agency has approved the seed. Seed certification implies that both the seed crop and seed lot have been duly inspected and that they meet requirements of good quality pedigree seeds.

Grow-Out Tests

Varieties being grown for seed production are periodically tested for genetic purity by grow-out tests, to make sure that they are being maintained in their true form.

Agronomic Principles

The seed production of varieties and hybrids should be carried out carefully in the region where these are well adopted. For example, seed production of cabbage is only possible in dry temperate areas where chilling requirements are met. The climatic factors have direct bearing on the quality seed production. Climate is the most important factor and generally for seed production, dry temperate climate is most suitable. Different vegetables need different climate for successful seed production and can be classified into temperate and tropical type. Climate may enhance bolting in the normal bulb crop of onion. Photoperiod also affects bulb crop and seed production in onion. Like climatic factors, the production and post harvest technologies will also play a major role in quality seed production. Hence, proper strategies to be planned for quality seed production especially in horticultural crops particularly in vegetable seed production.

Selection of Seed

The seed must be from authenticated seed source and appropriate class of seed must be used for further multiplication. *e.g.* For production of foundation seed breeder seed must be used.

Proof of purchase (bill) is necessary to identify the source along with the following details. Details on the tag *viz.*, name of the agency, purity, germination, validity period. Verify that all the bags are of the same variety, if packed in many bags.

Seed Treatments

Seed treatment should be made before sowing for speedy germination and quick establishment of seedling. Fungicides like Thiram or Captan may be used @ 2 g/kg of seed as dry dressing or slurry treatment (or) biocontrol agents like *Pseudomonas fluorescens* @ 10 g/kg (or) *Trichoderma viride* @ 4 g/kg can be used to check seed borne diseases. Treatment with biofertilizers like *Azospirillum* and *Rhizobium* should be given to the seeds of all crops as per recommendation for early vigorous growth. Pre-sowing seed treatment with chemicals, major and micronutrients and growth regulators increase the growth, flowering, seed set, yield, germinability, vigour and storability of seeds.

Sowing

Always sow the seeds in equal distance and the sowing depth should be 2.5 times of the seed size. Line sowing is advisable for seed crop. Adopt correct spacing then only we can achieve required population by providing equal opportunity to each plant to develop and mature which is not possible in broadcasted crop. The sowing of seed crops in rows helps in conducting effective plant protection measures, roguing operations and field inspections. In some crops (e.g. tomato, bhendi and gourds), exact ratios of male: female must be maintained for hybrid seed production.

During hybrid seed production, planting of two parents, namely female and male parent line has to be done in a definite proportion (eg. 6:1 or 8:2). If the hybrid seed production involves male sterile lines, border rows of the male parent may also be sown (or) transplanted.

Irrigation

Immediately after sowing irrigate the field if it is a direct sown crop, if transplanted, first irrigate the field and transplant the seedlings

For Transplantation

Raise nursery at proper time, protect the seedlings from damping off, and maintain required population per unit area to avoid damping off and competition between seedlings. Transplant the seedlings at 25 to 40 days after sowing depending upon the species. After sowing, irrigate the field at 3rd day as life irrigation and thereafter once in a week or 10 days interval depending on the water requirement of the seed crop.

Weeding

Apply herbicide as pre or post emergence, give one weeding if weed infestation is noticed.

Roguing

Very important operation to be carried out for the seed crop by the grower in order to maintain the purity of seed crop. It must be done periodically to keep the seed field free of rogues. The rogues may be of physical contaminants or genetic contaminants.

All the types of unwanted plants in seed field other than seed crop is called rogues.

Physical Contaminant Rogues

1. Weeds
2. Objectionable weed plants
3. Volunteer plants
4. Designated diseased plants
5. Other crop plants

Genetic Contaminants

1. Off types - Deviant of seed plant. Common for varieties and hybrids.
2. Pollen shedders - Presence of B line in A line in the case of hybrid seed production field.
3. Shedding tassel - Presence of tassel in the female row of maize hybrid seed production plot.
4. Selfed bolls - In the case of cotton hybrid seed production plot, pollination without emasculation and dusting.
5. Selfed fruits - In the case of vegetables.

General and Specific Standards

General Seed Certification standards

Deals with sources and classes of seed, phases of seed certification, seed standards for genetic purity, rejections, tags, labels and seals.

Specific Crop Standards (Tables 10.3 and 10.4)

a. Field standards
b. Seed standards

Table 10.3: Field Standards for Important Agricultural Crops

Crops	*Paddy*		*Maize*		*Sorghum*		*Bajra*		*Green-gram*		*Black-gram*		*Cowpea*		*Redgram*		*Sesame*		*Ground-nut*		*Castor*		*Cotton*	
	FS	*CS*	*FS*	*CS*	*FS*	*CS*	*FS*	*CS*	*FS*	*CS*	*FS*	*CS*	*FS*	*CS*	*FS*	*CS*	*FS*	*CS*	*FS*	*CS*	*FS*	*CS*	*FS*	*CS*
Off-type per cent	0.05	0.20	1.0	1.0	0.05	0.10	0.05	0.10	0.10	0.20	0.10	0.20	0.10	0.20	0.10	0.20	0.10	0.20	0.10	0.20	0.10	0.20	0.10	0.20
Inseparable other crop plants	-	-	-	-	-	-	-	-	-	-	_	_	_	_	-	-	-	-	-	-	_	_	_	_
Objectionable weed plants (per cent)	0.01	0.02	-	-	-	-	-	-	-	-	_	_	_	_	-	-	-	-	-	-	_	_	_	_
Plants or heads affected by designated diseases	-	-	-	-	0.05	-	0.05	0.10	0.10	0.20	_	_	0.10	0.20	-	-	0.50	1.0	-	-	_	_	_	_

Table 10.4: Field Standards for Important Horticultural Crops

Crops	*Tomato*		*Brinjal*		*Chilli*		*Bhendi*		*Cucurbits*		*Onion*	
	FS	*CS*	*FS*	*CS*	*FS*	*CS*	*FS*	*CS*	*FS*	*CS*	*FS*	*CS*
Off-type per cent	0.10	0.20	0.10	0.20	0.10	0.20	0.10	0.20	0.10	0.20	0.10	0.20
Inseparable other crop plants	–	–	–	–	–	–	–	–	–	–	–	–
Objectionable weed plants (per cent)	–	–	–	–	–	–	None	None	None	None	–	–
Plants or heads affected by designated diseases	0.10	0.50	0.10	0.50	0.10	0.50	–	–	–	–	–	–

a. Field standards

1. Land requirements
2. Minimum isolation requirements
3. Minimum specific crop standards for
   a. Off-types
   b. Diseases (designated diseases only)
   c. Objectionable weeds
   d. Inseparable other crop plants

Foliar Spray

To avoid or arrest flower shedding, foliar spray like NAA will be given to the seed field. *e.g.* Tomato and chilli, 2 sprays of 40 ppm of NAA at flowering will arrest the flower drop.

- In cotton, spraying of DAP or KCl at boll formation stage improves the seed quality.
- In pulses, spraying of 2 per cent DAP at flowering stage produces bolder seeds.

Plant Protection

Look for presence of designated diseases in the seed field. Remove the plants affected by designated diseases. The periodical plant protection measures will keep away the field from pest and diseases. Carefully select the pesticides and spray to the crop species which are pollinated by insects.

Harvesting

Harvest the seed crop at physiological maturity stage to get more seed yield and better quality of seed based on some visual indices. *e.g.* Sunflower - turning of yellow color of back side of thalamus is physiological maturity indices.

In the case of hybrids harvest the R line first, remove the produce from the field and do one final rouging in the female row and do the harvest.

Pre Storage Seed Treatments

Seed treatment with the fungicides and insecticides is done to arrest the carryover of pathogens and insects with the seed or their fresh entry into it.

The treatments given to the seeds prior to or during storage to protect the seed from deteriorative changes and from pest and diseases are called pre storage treatments. The types of pre - storage seed treatments are

i. Halogenation
ii. Antioxidant treatment
iii. Seed sanitation
iv. Seed fumigation

Table 10.5: Physiological and Harvestable Maturity Symptoms in different Crop Plants

Crop	*Attains Maximum Physiological Maturity Indices/Symptoms at*	*Harvestable Maturity Indices/Symptoms*
	CEREALS	
Paddy (*Oryza sativa*)	28 and 31 days after 50 per cent of spikelets in the panicle have flowered for medium and short duration varieties, respectively	Turning of 90 per cent of seed to straw or golden yellow colour and associates with moisture content of 20 per cent for short and medium duration varieties and 17 per cent for long duration varieties.
	MILLETS	
Sorghum (*Sorghum bicolor*)	Formation of black (Dunken) spot on the aleurone layer at 38 to 44 days after anthesis	The leaves turn yellow and present a dried up appearance. The seeds are hard and firm.
Maize (*Zea mays*)	Formation of milk line (the line on a germinal face of the maize seed) and black layer (appearance of a layer of black suberized cell) in parallel	Turning of sheath colour from green to yellow and dry at maturity. The seeds become fairly hard and dry.
African tall maize (*Zea sp*)	44 days after anthesis	-
Pearl millet (*Pennisetum glaucum*)	30 and 40 days after anthesis, 35 and 49 days after stigma emergence	The leaves turn yellow and present a dried appearance
	PULSES	
Blackgram (*Vigna mungo*)	42 days after anthesis with maximum seed dry weight	Leaves become dry and colour of the pods and seeds turn from green to black
Greengram (*Vigna radiata*)	25 days after anthesis with maximum fresh and dry weight of seed, high germination and vigour. Also pod colour changes from green to greenish yellow	Leaves will be dried and seeds turn into dark green and hard

Halogenation

Dry Dressing

It is application of halogens like chlorine, bromine and iodine to seeds before storage as dry dusting to stabilize the lipoprotein biomembrane.

Vapour Treatment

Exposure of seeds to halogen vapour in a closed container for 16-72 h, to very low concentrations of halogens. The concentration of chemicals and the time of exposure would depend on the material concerned. Halogens like chlorine, bromine and iodine or alcohols, such as ethanol, methanol or isopropanol are used.

Antioxidant Treatment

Application of antioxidants such as vitamin A, C, E, butylated hydroxytoluene (BHT) to seeds before storage as wet or dry treatments which delay the physiological ageing. The seeds are applied with antioxidants through soaking in respective solutions at particular concentration.

Seed Sanitation

The seed sanitation treatments refers to the application of pesticides (fungicides, insecticides or a combination of both) to seeds to disinfest and disinfect them from various seed borne and soil borne pathogenic organisms and storage insect pests to safe guard the seed and seedlings against the seed and soil borne pathogens. Seed borne pests and diseases may be carried within seeds or on seeds or they may accompany the seed as free living organisms or in debris. The seeds are simply dipped or soaked in the chemical solutions and the fungicides applied as dust, slurry or liquid in recommended concentrations.

Seed Fumigation

It is a process of exposing the seeds to fumigants (gaseous form of chemical) to control seed borne fungi and insects which cause seed deterioration during storage.

Mid Stroage Seed Treatments

Seeds in storage accumulate damage to cell membranes during senescence. Mid storage seed treatments are capable of reducing the age induced damages and restoring the seed vigour by controlling free radicals and consequent peroxidative damage to lipoprotein cell membranes. Besides, the seed viability and productivity of stored seeds are also improved.

Hydration - Dehydration

It is the process of soaking the low and medium vigour seeds in water with or without added chemicals usually for short durations to raise the seed moisture content to 25 – 30 per cent and drying back the seeds to safe limits for dry storage.

Packaging, Labeling and Sealing

It is necessary to use right type of containers with labeling and sealing in the prescribed manner for maintaining the seed moisture content at proper level.

Hybrid Seed Production Principles

All the operations are similar to the varieties except certain following steps.

Sowing

Follow the planting ratio recommended for that particular species.

Planting ratio is nothing but ratio to be maintained between the female and male lines in order to have good amount of pollen grains for proper pollination and fertilization. *e.g.* Planting ratio for sunflower hybrid seed production is 4:1. In the case of cotton it is called block system, the female and male are raised in a ratio of 8:2 in separate blocks not more than 5mt distance. This will aid in easy identification of the male plants and aids in pollen collection, since hybrid seed production is by emasculation and dusting.

Synchronized Flowering

Uniformity of flowering between parental lines. To achieve uniformity any one of techniques can be done

1. Staggered sowing
2. Withholding irrigation
3. Application of urea or DAP.

Chapter 11

Organic Farming: A Way Forward for Sustainable Agriculture

A. Bharani, E. Somasundaram and D. Udhaya Nandhini

Department of Sustainable Organic Agriculture,
Tamil Nadu Agricultural University,
Coimbatore – 641 003, Tamil Nadu

Introduction

Crop productivity and soil health are the deciding factor for country's strength. Presently the farming situation urges need to develop farming techniques, which are sustainable from environmental, production, and socio-economic points of view. Modern agricultural production throughout the world does not appear to be sustainable in the long run. Sustainable agricultural development is the management and conservation of the natural resource base and the orientation of technological and institutional change in such a manner to assure the attainment and continued satisfaction of human needs for the present and future generations. Such sustainable development in the agriculture, forestry and fishery sectors, conserves land, water, plant and animal genetic resources, is environmentally non-degrading, technically appropriate, economically viable and socially acceptable. Such concerns imparted a way to organic farming. It is the need of the day to understand the prospects and problems of organic farming to launch a successful and flawless organic production programme in the farm environment.

Organic Farming

It is the most widely recognized alternative farming system to chemical agriculture evolved during 1940's largely in response to the publication of J.I Rodale in the U.S. Lay Eve Bayour in England and Sir Albert Howard in India. In 1980, USDA released a landmark report on organic farming. The report defined organic farming as follows;

"Organic farming is a production system, which avoids or largely excludes the use of synthetically compounded fertilizers, pesticides, growth regulators, and livestock feed additives. To the maximum extent feasible, organic farming systems rely upon crop rotations, crop residues, animal manures, legumes, green manures, off-farm organic wastes, mechanical cultivation, mineral bearing rocks, and aspects of biological pest control to maintain soil productivity and tilth to supply plant nutrients, and to control insects, weeds, and other pests".

Organic agriculture system primarily aims at managing the agro- ecosystem as an autonomous system, based on the primary production capacity of the soil under the local conditions. It implies treating the system on any scale, as a living organism supporting its own vital potential for biomass and animal production, along with biological mechanisms for mineral balancing, soil improvement and pest and disease control. Organic farming in its holistic view refers to agriculture that aims to reflect the profound interrelationship that exists between farm biota its production and the overall environment (Scofield, 1986).

Organic agriculture is a unique production management system which promotes and enhances agro-ecosystem health, including biodiversity, biological cycles and soil biological activity, and this is accomplished by using on-farm agronomic, biological and mechanical methods in exclusion of all synthetic off-farm inputs.

Four principles of organic agriculture are:

1. **The Principle of Health** - Organic agriculture should sustain and enhance the health of soil, plant, animal and human as one and indivisible.
2. **The Principle of Ecology** - Organic agriculture should be based on living ecological systems and cycles, work with them, emulate them and help sustain them.
3. **The Principle of Fairness** - Organic agriculture should build on relationships that ensure fairness with regard to the common environment and life opportunities.
4. **The Principle of Care** - Organic agriculture should be managed in a precautionary and responsible manner to protect the health and well being of current and future generations and the environment.

These basic principles provide organic farming with a platform for ensuring the health of environment for sustainable development, even though the sustainable development of mankind is not directly specified in the principles.

Concepts of Organic Farming

1. Use of on- farm resources and avoiding/minimizing the use of off – farm resources.

2. Integrating farm resources for reducing the use of external inputs.
3. Soil health management using farm waste (crop weed and animal wastes).
4. Generating feed resources in the farm: by growing forage crops and fodder (shrub and trees).
5. Green biomass generation for soil fertility management by growing high foliage producing plants and trees.
6. Composting of crop residues and livestock manures: with the help of earth worms converting them into value added vermicomposts.
7. Preparing plant growth promoting and disease tolerant/controlling substances (like Panchagavya, Coconut milk slurry, Amirthakaraisal *etc.*).
8. Use of eco friendly pest management/crop protection measures (like use of predators and parasites, herbal extracts, crop diversification *etc.*).

Other Possible Aims are

1. To work as much as possible within a closed system and draw upon local resources.
2. To maintain the long term fertility of the soil.
3. To avoid all forms of pollution caused by agricultural techniques.
4. To provide a food stuff of high nutritional quality in sufficient quantity.
5. To reduce the use of fossil energy in agricultural practices to be minimum tending to zero.
6. To give to all livestock the conditions of life that conform to their physiological needs.
7. To make it possible for agricultural families to earn a living through their work and develop their potentialities as human beings.
8. To maintain the rural environment and also preserve non- agricultural ecological habitats.

How to Achieve Sustainability?

There are a few aspects to be strictly followed to achieve sustainability:

☆ **Enrichment of soil** – Abandon use of chemicals, use crop residue as mulch, use organic and biological fertilizers, adopt crop rotation and multiple cropping, avoid excessive tilling and keep soil covered with green cover or biological mulch.

☆ **Management of temperature** - Keep soil covered, Plant trees and bushes on bund.

☆ **Conservation of soil and rain water** – Dig percolation tanks, maintain contour bunds in sloppy land and adopt contour row cultivation, dig farm ponds, maintain low height plantation on bunds.

☆ **Harvesting of sun energy** – Maintain green stand throughout the year through combination of different crops and plantation schedules.

- **Self reliance in inputs** – develop your own seed, on-farm production of compost, vermicompost, vermiwash, liquid manures and botanical extracts.
- **Maintenance of life forms** – Develop habitat for sustenance of life forms, never use pesticides and create enough diversity.
- **Integration of animals** – Animals are important components of organic management and not only provide animal products but also provide enough dung and urine for use in soil.

What we need to do for sustainable organic agriculture?

- Rain water harvesting – ground water protection
- Planting of multipurpose trees
- Soil erosion reduction
- Animal husbandry
- Balancing the ecological balance
- Crop rotation

Potential Areas of Organic Farming

The organic farming perhaps is rarely practice and there is dearth of data on relative yield and financially viability with pure organic farming verses intensive farming. Whatever little organic farming that is being practiced in India seems to be taken up mainly as part of contract farming and it may be promoted in selected rainfed area and for export oriented crops in India (Marwaha and Jat 2004). Soil and climate conditions in India's rainfed areas make them particularly well suited to organic agriculture. Several states in India had already initiated steps to encourage organic farming among the farming community. Amongst the states, Madhya Pradesh took early lead in organic farming followed by Uttarakhand (declared as organic state) and others states. The 'Organic Farming Council of Punjab' was established in 2006 for the promotion of organic farming.

As far as potential areas are concerned three priority areas have been identified:

- **Category I:** Areas where fertilizers and other agrochemicals consumption is very less. These areas are in Assam and other north-eastern states, Jharkhand, Odisha, Jammu and Kashmir, Himachal Pradesh, Karnataka, Madhya Pradesh, Chhattisgarh and Rajasthan.
- **Category II:** Areas under rainfed and hilly areas offer tremendous opportunities for being converted into purely organic farming areas. Similarly, some area under tea, coffee, spices and cashew may be easily brought under organic farming with a trust on export of organic produce.
- **Category III:** Areas with irrigation and heavy use of fertilizers and other agrochemicals.

The appropriate strategy will be to promote organic agriculture in the areas under category I and II.

Strategies for Promoting Organic Farming

The strategy for promotion of organic farming identification of potential areas and crop is crucial. There is immense potential for promoting organic farming in India particularly in rainfed areas as regards crops, priority is for fruits, vegetables, spices, medicinal plants, oilseeds, pulses, cotton, wheat and basmati rice. Poor and marginal condition of soil as well as socio-economic condition of Indian farmers also support to promoting it (Meena *et al.*, 2013). For promoting organic farming in India we need to emphasise on the following area.

1. **Development of cropping system/intercropping/mixed farming based on organic inputs/resources**

Developing of suitable organic farming systems and practices in different agro-climatic and farming situations. There should be region specific particularly in rainfed areas, because in rainfed areas there is limited scope for chemical fertilizers. Development of package of practices for various cropping system/intercropping/ mixed cropping with high value crops based on organic inputs.

2. **Technology for promotion of organic farming**

Recent development technologies like efficient use of crop residues, recycling of residues of dual purpose legume, use of biomass of some of the non conventional shrubs and trees, biogas slurry and vermicompost for nutrient supply and use of bio-agents and predators for controlling pest and disease make organic farming a comfortable promotion in selected areas and high value crops.

3. **Accreditation of certified agencies**

Certification is a process of labeling in term that denote products have been produced in accordance with standards during food production, handling, processing and marketing stages and certified duly by a certification body. Certification and regulatory mechanism intended to assure quality and prevent fraud. For organic producers, certification identifies suppliers of products approved for use in certified operations. At present there are 12 accredited certifying agencies in various places in the country.

Options for Organic Agriculture in Rainfed Areas

Livestock

Profit of rain fed areas and to achieve the objectives of organic farming is mainly depends on the livestock. The easy for multiplying the farm income are cow and goat rearing. Things we receive from dairy are the main components in organic inputs.

Soil Fertility

A great emphasis is placed to maintain the soil fertility by returning all the wastes to it chiefly through compost to minimize the gap between NPK addition and removal from the soil (Chhonkar, 2002). Biofertilisers, green manures, sheep penning, and green leaf manures are used to improve the soil fertility. All these technologies are possible in rainfed situation. Nitrogen fixing trees can be planted along the borders of the field ensuring the availability of green leaf manures. These trees are not only fixing atmospheric nitrogen but also used for feed, fuel and woody purpose. These leaf clippings can be put into manure pits and made as compost. In addition, trees can be used as biofencing and windbreaks that will increase water filtration.

Crop Rotation

Recurrent succession of crop on the same piece of land either in a year or over a longer period of time is called crop rotation. Component crops are so chosen such that soil health is not impaired (*e.g.*) cotton-gram; sugarcane-wheat. Otherwise it means growing a set of crop in a regular succession on a piece of land in a specific period of time, with an object to get maximum profit out of least investment without impairing soil fertility. Soil health will be improved by crop rotation. Short duration crops are best suitable for crop rotation.

Soil Health

Report says that soil microbial population is increased in organic agriculture. Organic agriculture enhances the activity of soil, plant and animal.

How can Water Crisis be Tackled in Organic Agriculture?

Modern agriculture technologies are mainly depending on irrigation water. So, continuous and higher amount of water is required for modern agriculture. Soil organic matter acts as a source for mineral nutrients resulting in aggregate formation, due to which soil erosion gets reduced.

The Global World of Organic Agriculture - 2010

As per the details released at BioFach 2010 at Nuremberg, organic agriculture is developing rapidly, and statistical information is now available from 160 countries of the world. Its share of agricultural land and farms continues to grow in many countries. The main results of the latest global survey on certified organic farming are summarized below:

Growing Area under Certified Organic Agriculture

- About 35 million hectares of agricultural land are managed organically by almost 1.4 million producers.
- The regions with the largest areas of organically managed agricultural land are Oceania (12.1 million hectares), Europe (8.2 million hectares) and Latin America (8.1 million hectares). The countries with the most organic agricultural land are Australia, Argentina and China. The highest shares of organically managed agricultural land are in the Falkland Islands (36.9 per cent), Liechtenstein (29.8 per cent) and Austria (15.9 per cent).
- The countries with the highest numbers of producers are India (340'000 producers), Uganda (180'000) and Mexico (130'000). More than one third of organic producers are in Africa.

- ☆ On a global level, the organic agricultural land area increased in all regions, in total by almost three million hectares, or nine per cent, compared to the data from 2007.
- ☆ Twenty-six per cent (or 1.65 million hectares) more land under organic management was reported in Latin America, mainly due to strong growth in Argentina. In Europe the organic agricultural land increased by more than half a million hectares, in Asia by 0.4 million.
- ☆ About one-third of the world's organically managed agricultural land (i.e) 12 million hectares is located in developing countries. Most of this land is in Latin America, with Asia and Africa in second and third place. The countries with the largest area under organic management are Argentina, China and Brazil.
- ☆ About 31 million hectares are organic wild collection areas and land for bee keeping. The majority of this land is in developing countries contrast to agricultural land, of which two-thirds is in developed countries. Further, organic areas include aquaculture areas (0.43 million hectares), forest (0.01 million hectares) and grazed non-agricultural land (0.32 million hectares).

Organic Agriculture in India

The organic movement in India has its origin in the work of Albert Howard (1940) who formulated and conceptualized most of the views which were later accepted by those people who became active in this movement.

The key characteristics include protecting the long term fertility of soils by maintaining organic matter levels, fostering soil biological activity, careful mechanical intervention, nitrogen self sufficiency through the use of legumes and biological nitrogen fixation, effective recycling of organic materials including crop residues and livestock wastes and weed and diseases and pest control relying primarily on crop rotations, natural predators, diversity, organic manuring and resistant varieties (Chandra and Chauhan, 2004).

Since January 1994 "Sevagram Declaration" for promotion of organic agriculture in India, organic farming has grown many folds and number of initiatives at Government and Non-Government level has given it a firm direction. While National Programme on Organic Production (NPOP) defined its regulatory framework, the National Project on Organic Farming (NPOF) has defined the promotion strategy and provided necessary support for area expansion under certified organic farming.

Growing Certified Area

Before the implementation of NPOP during 2001 and introduction of accreditation process for certification agencies, there was no institutional arrangement for assessment of organically certified area. Initial estimates during 2003-04 suggested that approximately 42,000 ha of cultivated land were certified organic. By 2009, India had brought more than 9.2 million ha of land under certification. Out of this while cultivable land was approximately 1.2 million ha, remaining 8 million ha area was forest land for wild collection. Growing awareness, increasing market demand, increasing inclination of farmers to go organic and growing institutional support has resulted into phenomenal growth in total certified area during the last five years. As on March 2009, total area under organic certification process stood at 12.01 lakh ha (Yadav *et al.*, 2013).

Important Features of Indian Organic Sector

The key factors affecting consumer demand for organic food is the health consciousness and the key willingness of the public to pay for the high-priced produce. In general, consumers of organic products are an affluent, educated and health conscious group spurred by strong consumer demand, generous price premium and concern about the environment. Because of these hidden benefits, conventional growers are turning to organic farming (Sofia *et al.*, 2006).

With the phenomenal growth in area under organic management and growing demand for wild harvest products India has emerged as the single largest country with highest arable cultivated land under organic management. India has also achieved the status of single largest country in terms of total area under certified organic wild harvest collection. With the production of more than 77,000 MT of organic cotton lint India had achieved the status of largest organic cotton grower in the world a year ago, with more than 50 per cent of total world's organic cotton.

Growing Organic Food Market

Although no systematic information is available on size of organic food market as surveyed by the International Competence Centre for Organic Agriculture (ICCOA) in top 8 metro cities of India (which comprise about 5.3 per cent of the households), the market potential for organic foods in 2006 in top 8 metros of the country was at ₹ 562 crore taking into account current purchase patterns of consumer in modern retail format. The overall market potential is estimated to be around ₹1452 crores.

Future Prospects

The movement started with developed world, gradually picking up in developing countries. But demand is still concentrated in developed and most affluent countries. Local demand for organic food is growing. India is poised for faster growth with growing domestic market. Success of organic movement in India depends upon the growth of its own domestic markets. India has traditionally been a country of organic agriculture, but the growth of modern scientific, input intensive agriculture has pushed it to wall. But with the increasing awareness about the safety and quality of foods, long term sustainability of the system and accumulating evidences of being equally productive, the organic farming has emerged as an alternative system of farming which not only addresses the quality and sustainability concerns, but also ensures a debt free, profitable livelihood option.

Conclusion

Organic farming can provide quality food without adversely affecting the soils health and the environment. There is need to identify suitable crops/products on regional basis for organic production with sustainable resource utilization within the farm to attain sustainability.

References

Albert Howard. 1940. An Agricultural Testament, Oxford University Press,

Chandra, S and S. K. Chauhan. 2004. "Prospects of organic farming in India," Indian Farming. 52(2): 11–14.

Chhonkar, P.K. 2002. Organic farming myth and reality, in Proceedings of the FAI Seminar on Fertilizer and Agriculture Meeting the Challenges, New Delhi, India, December 2002.

Marwaha, B.C and S.L. Jat. 2004. Statistics and scope of organic farming in India. Fert. News: 49(11): 41-48.

Meena, B.P., V. D. Meena, M. L. Dotania and N. K. Sinha. 2013. Potential Areas and Strategies for Promoting of Organic Farming. Popular Kheti 1(4): 32-37

Scofield, A, 1986. Organic farming –the origin of the name. Biological Agriculture and Horticulture, 4: 1-5.

Sofia, P. K., R. Prasad, and V. K. Vijay. 2006. "Organic farming-tradition reinvented," Indian Journal of Traditional Knowledge, 5(1): 139–142.

Yadav, S.K. Subhash Babu, M. K. Yadav, Kalyan Singh, G. S. Yadav, and Suresh Pal. 2013. A Review of Organic Farming for Sustainable Agriculture in Northern India. International Journal of Agronomy.

www.agritec.com

www.apeda.gov.in

www.ciks.org

www.cpreec.org

www.ncipm.org.in

www.sciencedirect.com

www.sristi.org

www.tnau.ac.in

ecological basis for organic production with sustainable resource utilization within the farm to attain sustainability.

References

Albert Howard. 1940. *An Agricultural Testament*. Oxford University Press.

Chhonkar, P.K. and S.K. Chauhan. 2004. Prospects of organic farming in India. *Indian Farming*. 54(2): 11-14.

Chandra, D.K. 2012. Organic farming myth and reality. In Proceedings of the IAI Seminar on Agriculture: Meeting the Challenges, New Delhi, India, [illegible] 2012.

Nandan [illegible] India [illegible]

Chapter 12

Weather Based Crop Advisory Prescriptions for Climate Smart Agriculture

S. Panneerselvam

Professor and Head,
Agro Climate Research Centre,
Tamil Nadu Agricultural University,
Coimbatore – 641 003, Tamil Nadu

Agriculture is the pulse of our country and by and large forms the backbone of national economy. The demand for agriculture output in India will grow exponentially over coming decades due to population growth and expanding Indian economies. To meet this expanding demand, agriculture must become more adept at anticipating climate changes and variations and finding ways of adapting to these changes. However, in order to keep pace with the increasing population, the growth in agricultural production should be sustainable in the long run. Weather plays an important role in agricultural production. It has a profound influence on crop growth, development and yields; on the incidence of pests and diseases; on water needs; and on fertilizer requirements. Weather aberrations may cause physical damage to crops and soil erosion. The quality of crop produce during movement from field to storage and transport to market depends on weather. Bad weather may affect the quality of produce during transport, and the viability and vigour of seeds and planting material during storage. Thus, there is no aspect of crop culture that is immune to the impact of weather.

Weather and climate information has the potential to reduce the impact of adverse weather events. This will occur because the advance notice will allow decision makers the opportunity to implement plans to minimize the impact of adverse events and find opportunities within favorable events. New advances in computing and modelling have resulted in predicting the climate/weather satisfactorily to a great extent. Such weather and climate information in advance is able to reduce uncertainty and risk of the farmers and helps in better decision making. It leads to better economic, social and environmental outcome for farmers.

For optimal productivity at a given location, crops and cropping practices must be such that while their cardinal phased weather requirements match the temporal march of the relevant weather element(s), endemic periods of pests, diseases and hazardous weather are avoided. In such strategic planning of crops and cropping practices, short-period climatic data, both routine and processed (such as initial and conditional probabilities), have a vital role to play.

Weather Forecast

Weather and climatic information plays a major role before and during the cropping season. Occurrences of erratic weather are beyond human control. It is possible, however, to adapt to or mitigate the effects of adverse weather if a forecast of the expected weather is provided in advance to the farmer to organize and activate their own resources in order to reap the benefits. Agronomic strategies to cope with changing weather are available. For example, delays in the start of crop season can be countered by using short-duration varieties or crops and thicker sowings. Once the crop season starts, however, the resources and technology get committed and the only option left then is to adopt crop-cultural practices to minimize the effects of mid-seasonal hazardous weather phenomena, while relying on advance notice of their occurrence. For example, resorting to irrigation or lighting trash fires can prevent the effects of frosts. Thus, medium range weather forecasts with a validity period that enables farmers to organize and carry out appropriate cultural operations to cope with or take advantage of the forecasted weather are clearly useful. The rapid advances in information technology and its spread to rural areas provide better opportunities to meet the rising demand among farmers for timely and accurate weather forecasts.

Agro-meteorological Agro Advisory Prescriptions

Agricultural crops are primarily influenced by weather and Indian agriculture is mostly governed by the monsoon rainfall in spite of extensive technological developments, improved irrigation facilities *etc.* Thus the weather forecast presume significant role for planning and managing agricultural activities over the region. Observing and understanding weather for better management of agricultural crops was emphasized by Kautilya, way back in 3rd century B.C. The invention of instruments like barometer by Evangelista Torricilli for measurement of atmospheric pressure in 1643 laid the foundation for development of scientific methods of observing and forecasting weather. In India collection of weather observations were started after the establishment of India Meteorological Department (IMD) in 1875. India Meteorological Department (IMD) started weather services for farmers in the year 1945. It was broadcast by All India Radio in the form of Farmer's Weather Bulletin (FWB). Subsequently, in the year 1976, IMD started Agro-Meteorological agricultural Advisory Service (AAS) from its State Meteorological Centers, in collaboration with Agriculture Departments of the respective State Governments. Though these services are being regularly provided by IMD for the past many years, the demand of the farming community could not be fully met due to certain drawbacks in the system. In view of that IMD launched Integrated Agromet Service in the country for 2007 in collaboration with different organizations/institutes.

National Agromet Advisory Bulletin

The bulletin is prepared for national level agricultural-planning and management and is being issued by National Agromet Advisory Service Centre, Agricultural Meteorology Division, India Meteorological Department. Prime users of this bulletin are Crop Weather Watch Group, (CWWG), Ministry of agriculture. Bulletin is also communicated to all the related Ministries (State and Central), Organizations, NGOs for their use.

State Agromet Advisory Bulletin

This bulletin is prepared for State level agricultural planning and management. These bulletins are issued from 22 AAS units at different State capitals. Prime user of this bulletin is State ACWWG. This is also meant for other users like Fertilizer industry, Pesticide industry, Irrigation Department, Seed Corporation, Transport and other organizations which provide inputs in agriculture.

District Agromet Advisory Bulletin

This is prepared for the farmers of the districts. These bulletins are being issued from 30 AMFUs functioning at State Agricultural Universities. This contains advisories for all the weather sensitive agricultural operations form sowing to harvest. It also includes advisories for horticultural crops and livestock. These weather based advisories are disseminated to the farmers through mass media dissemination, Internet etc as well as through district level intermediaries. The advisories will be communicated through multi-channel dissemination system.

Agro Meteorological Services

Activities in agro meteorology in India started with the setting up of division of Agro meteorology in India Meteorological Department (IMO) in the year 1932. To collect meteorological data, the IMD has 35 meteorological subdivisions spread all over the country. Special weather forecast for farmers is being issued by IMD through its regional network of stations. The forecasts issued by it are mostly general in nature and not specific to any particular area or to any crop. Hence, these are of not much beneficial to the farmers. Realizing the need for medium range weather forecasting, a National Centre for Medium Range Weather Forecasting (NCMRWF) established by the Department of Science and Technology in 1988. Its objectives include development of operational medium range (3 -10 days in advance) weather forecasting capabilities and Agrometeorology Advisory service for 127 agro climatic zones. Till date, 112 centres have been established in State Agricultural University's (SAUs) research stations. Medium range weather forecast issued by NCMRWF once/

twice a week, covering a period of 3 to 5 days, are utilized by Agrometeorological Field Units (AMFU) located in SAUs for preparing weather-based agro advisories. From June 2008 the IMD has reorganized and started issuing district level forecast now generated by respective Regional Meteorological Centre. The district level forecast being issued with a lead time of 5 days.

The IMD has established Agromet Advisory Service Units (AASUs) at the meteorological offices of the state headquarters. These AASUs issue bi-weekly agromet advisories to the states. First, the condition of the crops in the state and then advisory on the farming operations, based on the past weather/likely future weather realized is provided. The rainfall forecasts valid for the next two days and the outlook valid for two subsequent days are also given. Assistance for the agricultural related aspects is taken from the state agricultural universities and agricultural departments.

Format of Agro Advisory

The advisory must consist of weather summary of the last week, climatic normals, weather forecasts and crop moisture index, drought severity index, soil moisture status *etc.* it should also provide information on phenological stages of the crop, pest and diseases and crop stress conditions. The advisory content may include crop management practices such as time of sowing, likely date of next irrigation, control operations of pests and diseases, fertilizer applications *etc.*, based on the likely situation of weather in the coming few day. Information on crop planning, variety selection, harvesting time may also be provided. It may also have information on spraying conditions for insect, weeds *etc.*, animal health, and live stock management with respect to housing, health and nutrition. The IMD also prepares crop weather calendars for different crops. They contain information on weather happenings growth phase-wise for different crops. Adverse weather conditions are also mentioned. These calendars are prepared in consultation with State Departments of Agriculture. These calendars serve as guide for agricultural operations. The advisories should serve as an early warning function for extreme weather events such as heavy rain, flash floods, strong winds and temperatures.

Essentials of Weather Based Agro Meteorologigal Services

Any extension service has its own essentials in order to effectively pass on the information to the recipient concern. The weather based Agro Advisory service is not an exception as we are dealing with farmers who has different knowledge level and has varying degree of accessibility to the knowledge being imparted to them. Though the media and medium of communication differ with individuals the core requirement that one should essentially know is important for such knowledge sharing activities. Hence, one should have the basic information about the area in which they are working as detailed below.

Real Time Information on the Crops

- ✰ Major crops grown in target area
- ✰ Varieties need to be grown

- ☆ Normal yield expected for early variety
- ☆ Time of seeding
- ☆ Phenological stage of the crop
- ☆ Crop Phenological specific pest and diseases.

Identification of Weather Sensitive Farm Operations

- ☆ Field preparation (Summer ploughing)
- ☆ Time of seeding
- ☆ Emergence of seedlings (crust formation-apply light irrigation if no rain forecast)
- ☆ Fertilizer application
- ☆ Irrigation
- ☆ Spraying of pesticides
- ☆ Outbreak of pests and diseases
- ☆ Harvesting
- ☆ Marketing

Accurate Weather Forecast

The above field operations are influenced by various weather parameters which need to be forecasted for developing agro advisories.

- ☆ Temperatures (maximum and minimum)
- ☆ Rainfall
- ☆ Wind direction
- ☆ Wind speed and
- ☆ Cloud cover

Reliable Source of Information

- ☆ Field visit by Agricultural officers of the state department
- ☆ Diagnostic field visits constituted by the university

Preparation of Crop Weather Calendars

- ☆ There is a need for structuring crop weather calendars depending upon the date of sowing and variety with major emphasis on vulnerability of the crops to the changing weather conditions.
- ☆ These calendars should specify the forecasting needs in terms of weather parameters and days in advance so that the forecast will be more appropriate. Ex: - Sudden rise in maximum temperature from 2.3 to 5.0°C favour the incidence of groundnut leaf miner at 35 days after sowing.

- ☆ Modeling approach to predict the anticipated changes in the crop condition based on weather forecast.
- ☆ There may be periods during which crops are not grown.

Easily Understandable Language

- ☆ Forecast should be in local language
- ☆ Advisory also be in local language and clearly understandable to the farmers.

The Advisory Should Contain

- ☆ Weather forecasts are issued by the meteorological services
- ☆ Anticipated definite changes on the crop condition
- ☆ Technologies recommended

Quick Dissemination of Information to Farmers Through

- ☆ TV channels
- ☆ Exclusive regional TV channel for weather and agro met advisories
- ☆ All India Radio
- ☆ Local Newspaper
- ☆ Messages directly to the fanners cell phone from the computer
- ☆ Agricultural extension officers of state department of Agriculture.

Communication Activities Required

- ☆ KVK's, Farmer's clubs and NGO's need to be involved in the dissemination of advisories.

Government Interventions

- ☆ Training programmes need to be conducted on the use of agromet information and advisories.

Benefits from Agro Met Services

The benefits bestowed to the farmers are documented and their' feedback is utilized to improve the advisory system. This will help in making the advisory system dynamic and subjected to the constant improvement in the overall up-liftment of the farmer through better planning and execution of the farmer's field operations.

Weather based Prescriptions

Field Preparation

- ☆ Field preparation is influenced by summer showers and high temperatures

- ☆ Summer ploughing helps in destruction of hibernating larvae and resistant spores of the pathogens thereby less attack by pests and diseases during forthcoming season.
- ☆ Timely sowing of rainfed crops
- ☆ More moisture conservation
- ☆ Control the weeds

Time of Seeding

This is influenced by rainfall.

- ☆ Rain forecast helps timely sowing of rainfed crops thereby better emergence and stand establishment.
- ☆ Later sowing (after July 15th) results in occurrence of yellow mosaic disease in black gram.
- ☆ Early sowing (June) of pigeon pea results in attack of wilt disease.
- ☆ Late sowing (after July) of pigeon pea results in incidence of *Phytophthora* blight
- ☆ Early sowing (before October 15th) favours the incidence of wilt in chickpea due to high temperatures.

Emergence of Seedlings

It is influenced by rainfall and high wind speed.

- ☆ Dry spell due to no rain forecast results in crust formation
- ☆ Light irrigations help in better seedling emergence.
- ☆ Prolonged dry spells results in dry root rot in groundnut.

Fertilizer Application

- ☆ Optimum moisture and nutrient availability helps for better growth and yield of crops.
- ☆ Rain forecasts helps in timely fertilizer application and thereby increase in fertilizer use efficiency.
- ☆ Fertilizer application during no rain forecast day's results in economic loss.

Irrigation

This operation is influenced by rainfall, temperature and wind speed.

- ☆ Irrigating the field during no rain forecast days helps in better growth
- ☆ Avoiding of irrigation during rain forecast days saves in irrigation water as well as tariff on energy.
- ☆ High wind speed results in high evaporation compelling the farmers to irrigate

- ☆ Irrigation and rainfall occurrence during rain forecast results in water logging in the fields there by resulting the yellowing and stunting of crop.
- ☆ Heat wave conditions: Application of light irrigations during evening hours protects the perennial crops.
- ☆ Cool wave conditions: Irrigating the fields during evening hours warm up the soil temperature thereby resulting in enhanced physiological activity of the crop.

Outbreak of Pests and Diseases

It is influenced by temperature, rainfall, relative humidity and cloud cover.

Favourable Conditions for Outbreak Leaf Miner in Groundnut

- ☆ Sudden raise in the maximum temperature from 2.3 to 5°C
- ☆ Minimum temperature of 20- 22°C
- ☆ Relative humidity more than 80 per cent
- ☆ Rainfall including traces for three consecutive days
- ☆ Sun shine hours of less than five hours a day.

Favourable Conditions for Outbreak of Rice Blast

- ☆ High humidity (more than 80 per cent) and minimum temperature (16 to 19°C) or cloudy sky with rainfall continuously for 5-6 days.
- ☆ Prolonged dry spell conditions favours multiplication of sucking pest complex on different crops.

Harvesting

Early harvest due to rain forecast helps in preventing the pre-harvest sprouting.

Special Agricultural Weather Prescriptions

Special agricultural weather prescriptions provide the necessary meteorological information to aid farmers in making certain special "crop- and/or cost-saving" decisions regarding farm operations. For the same temporal distribution of weather parameters, different crops will react differently.

Special forecasts are normally issued every day for a specific operation and generally cover the next 12–24 hours, with a further outlook if necessary. These special weather forecasts must be written by a trained agricultural meteorologist in consultation with farm management specialists for the current problems. They are normally issued for planting, irrigation, applying agricultural chemicals, cultivation, harvest and post-harvest processing, and they may also address other weather-related agricultural problems associated with the crop, its stage and location. Temperature bulletins for protection against freezing are also issued as special forecasts in areas where crops may suffer damage from freezing.

Field Preparation

Field preparation for rainfed crops is weather- dependent. In any dry land areas the amount of rainfall is very meager and farmers should take advantage of even minimum showers. Otherwise the moisture is lost. Minimal tillage is the current agronomic mantra for conserving moisture, retaining nutrients and keeping weeds out. An optimum soil moisture profile characterized by top dry soil, sub-surface moist soil and wet soil in the seeding zone is required to carry out field preparation for dry land farming. The prediction of the exact time of occurrence of rainfall in a particular location helps to initiate field preparation. An example of this is: "Pre-monsoon showers are expected in the 37th standard week of this year and farmers are requested to initiate field preparation activities before this week".

Sowing/Planting

Seed germination is dependent upon proper light and moisture, as well as soil temperature. Even with no nutritional or soil moisture constraints, root foraging capacities vary among crops in the same soil and within the same crop in different soils. Alternating temperatures assist the germination of many species of seeds and do not unfavourably affect the germination of those that do well under constant temperatures. The amplitude, which is the difference between maximum and minimum temperatures, decreases with depth and becomes negligible at a depth of 30 cm. Hence soil temperatures at 30 cm can be taken as constant. The temperature range at which soil temperatures will equal the air temperature will principally depend on the texture and structure of soil. Under a ground- shading crop, the depth of no diurnal change is pushed up compared to that seen under bare soil.

Application of Agricultural Chemicals

Use of agricultural chemicals is inevitable in crop production. Overuse of agrochemicals such as fungicides and pesticides, however, especially the systemic types and inorganic nitrogenous fertilizers, leads to contamination of food produce and soil; pollution of air, aquifers and water reservoirs; and development of chemo resistant strains of pests, diseases and weeds. Weather forecasts as detailed in the ensuing sections on control of insects, diseases and weeds can not only help minimize the volume of agrochemicals applied, but also make the applications more effective.

Soil Application

Precipitation is the most important factor that determines the efficacy of chemicals applied through soil. Precipitation in the succeeding 24 hours is the critical parameter. Limiting the amount of treatment through the effective use of weather information also leads to minimum pollution of groundwater and runoff.

Examples of forecasts for application of agricultural chemicals are:

Wind speeds are expected to be mostly favourable for application of agricultural chemicals today and tomorrow. Wind direction will be variable and wind speed will range from 6 to13 km/h in the forenoon and will become southerly with speeds of 13 to 24 km/h during the late afternoon. Temperatures are likely to exceed 27°C tomorrow. So caution should be exercised in applying oil-based sprays.

Heavy rain is expected in the next 24 hours, so foliar application of chemicals may be postponed.

Evaporation Losses for Irrigation

Irrigation water is costly to farmers in most agro-ecosystems today. Overuse can be both expensive and detrimental to the crop, while underuse can result in loss of crop quantity as well as quality. Estimates of daily consumptive use can be related to the free water loss from a Class A-type evaporation pan: the free water loss over the previous day for an area is obtained from the actual values recorded, while the loss for the succeeding 24 hours must be forecast based on the forecasts of rain, wind, relative humidity and bright hours of sunshine. For example, due to wetting of its surroundings by rain, the evaporation from a pan can be 20 to 30 per cent lower than with dry surroundings. Linear approximations have been derived for the estimation of solar radiation from bright hours of sunshine, potential evapotranspiration from either pan evaporation or from associated wind, and vapour pressure deficit terms.

The following are examples of water-loss forecasts:

Free water loss during the past 24 hours averaged 0.6 cm. Expected free water loss is 0.6 cm today and 0.8 cm tomorrow. Rainfall probability will remain low for the remainder of the week and crops will begin to suffer from moisture stress in four days' time. Supplementary irrigation of 7 cm in two days' time is recommended.

Rain is likely to occur in the next 24 hours in most of the areas in this region and so farmers may postpone their irrigation for this period.

Conclusion

The weather has a direct impact on the lives and livelihoods of the people throughout the world. Improving the flow of weather prescriptions to the public and the policy makers can improve day-to-day decision-making and influence critical decisions in agriculture and also in many sectors of the national economies.

References

Boyd, C.E. and C.S. Tucker, 1998: *Pond Aquaculture Water Quality Management.* Boston, Kluwer Academic Publishers.

Cushing, D.H., 1982: *Climate and Fisheries.* London, Academic Press.

Glantz, M.H. and L.E. Feingold, 1992: Climatic variability, climate change, and fisheries: a summary. In: *Climate Variability, Climate Change and Fisheries* (M.H. Glantz and L.E. Feingold, eds). London, Cambridge University Press.

Gommes, R., H.P. Das, L. Mariani, A. Challinor, B. Tychon, R. Balaghi and M.A.A. Dawod, 2007. WMO Guide to Agrometeorological Practices, Chapter 5, Agrometeorological Forecasting.70pp.http: //www.agrometeorology.org/ fileadmin/insam/repository/gamp_chapt5.pdf.

Intergovernmental Panel on Climate Change (IPCC), 2001: *Climate Change 2001: Impacts, Adaptations and Vulnerability. Contribution of Working Group II to the Third Assessment Report of the IPCC.* Cambridge, Cambridge University Press.

Chapter 13

Promising Bioinoculants to different Crops

P. Marimuthu, N. Jaivel and R. Rajesh

Department of Agricultural Microbiology,
Tamil Nadu Agricultural University,
Coimbatore – 641 003, Tamil Nadu

Biofertilizers are defined as preparations containing living cells or latent cells of efficient strains of microorganisms that help crop plants' uptake of nutrients by their interactions in the rhizosphere when applied through seed or soil. They accelerate certain microbial processes in the soil which augment the extent of availability of nutrients in a form easily assimilated by plants. Very often microorganisms are not as efficient in natural surroundings as one would expect them to be and therefore artificially multiplied cultures of efficient selected microorganisms play a vital role in accelerating the microbial processes in soil.

Use of biofertilizers is one of the important components of integrated nutrient management, as they are cost effective and renewable source of plant nutrients to supplement the chemical fertilizers for sustainable agriculture. Several microorganisms and their association with crop plants are being exploited in the production of biofertilizers. The nitrogen fixing bacteria, *Rhizobium*, *Azospirillum*, *Azotobacter*, *Gluconacetobacter*, *cyanobacteria*; PO_4 solubilizing bacteria, P mobilizing fungi, Arbuscular mycorrhiza and N fixing green manure, Azolla are presently multiplied in large quantity and distributed to the farmers.

The production of food grain is required to be increased continuously keeping in view of the increased growth in population and in this context the requirement of inorganic fertilizer will also be in the ascending order. Keeping in mind, the work on biofertilizers, the suitable alternate supplement for inorganic fertilizers has been activated in the country. Although the production of *Rhizobium* was started in 1934, its first commercial production was initiated in 1956 at Indian Agricultural Research Institute, New Delhi and Agricultural college and Research Institute (now Tamil Nadu Agricultural University), Coimbatore. This was followed by the production of different types of biofertilizers in smaller quantities in State Agricultural Universities. Some state governments of the provinces have also started production units to meet their local demand. The research activities on BNF have also strengthened in Agricultural Universities. In 1978, the Indian Council of Agricultural Research (ICAR) has started an All India Coordinated Research Project on BNF. Subsequently, the Government of India has launched a National Project on Development and Use of Biofertilizers in 1983. Later in 1988-89, the Department of Biotechnology, Ministry of Science and Technology, Government of India initiated massive project on Technology Development and Demonstration of Biofertilizers for rice crop. Currently, more than 95 firms belonging to public and private sectors are involved in production of different types of biofertilizers at different parts of the country.

The well-established and resourceful biofertilizers like *Rhizobium*, dominated the production earlier, but currently, other microbial inoculants like *Azotobacter*, *Azospirillum*, *Acetobacter* and phosphate solubilizing microorganisms (PSM) are being commercially produced and marketed. Recently, the production of arbuscular mycorrhizae (VAM) has also been initiated in some production centers. Fertilizer Association of India has reported that the production of biofertilizers in India increased from 2005 mt in 1992-93 to 10000 mt in 1999-2000 (FAI, 2001). The current production of bioinoculant in the country is 20,040.35 tonnes annually with a annual growth rate of about 5-6 per cent. Still the full installed capacity has not been fully utilized due to varied reasons including several constraints (Kannaiyan *et al.*, 2003). The installed biofertilizer (*Rhizobium, Azotobacter, Azospirillum* and P solubilizers) production capacity of about 20,000 tonnes is much lower than the potential demand, which is estimated to be 3.4 lakh metric tonnes as observed by Biotech Consortium India Ltd., Delhi (Tandon, 2009). This indicated substantial growth of this microbial industry and eventually entrepreneurship.

Although farmers are willing to try out the use of inoculant, a failure to observe any benefits due to the ineffectiveness of the inoculant causes the farmer to be reluctant to make the extra efforts and expenses to repeat its use. Therefore to promote the inoculant utilization in a country, especially in a developing country, it is very necessary to introduce high quality inoculant to farmers along with proper recommendation for the first use of inoculant to ensure that a good response of legumes to inoculation is achieved. Hence it is our duty to supply high-performing, good quality, timely available inoculants to the crops to ensure the maximum endurance of beneficial effects from the bioinoculants.

Continuous application of chemical fertilizers for the increased yield leads to reduction in soil beneficial microbial population and nutrient level and ultimately the fertility of the soil. Hence, application of biofertilizer in addition to the recommended dose of fertilizer results in increased nutrient use efficiency and at the same time it is cost effective and ecofriendly also. Merely applying chemical fertilizers could not increase the crop yield since it is not readily available for crop assimilation and hence results in yield reduction. Chemical fertilizers along with biofertilizers result in prolonged availability and assimilation of nutrients by crop plants.

In addition, the use of these biofertilizers increases crop yield by 20-30 per cent and replaces chemical fertilizers by 25 per cent. Biofertilizers are available for almost all crops with crop specificity and soil specificity.

Rhizobium

Rhizobium is one of the nitrogen fixing biofertilizers applied for pulse crops. Different crop specific strains are *Rhizobium* available for effective nodulation and nitrogen fixation. The other biofertilizers used for pulses are Phosphorus solubilizers (PSB) and Phosphorus mobilizers and PGPR (Pseudomonas). They exhibit or work together in a synergistic manner, when applied together and ultimately resulted in higher yields. In general, the trials carried out in pulses, with their specific strains had resulted in an average yield increase from 11- 23 per cent from the uninoculated control.

Azospirillum

The only versatile and omnipotent nitrogen fixing organism can be used for almost all types of crops except pulses. Mostly this is associated with the roots of cereals and millets; it is an associative symbiont. In addition to nitrogen fixation, it also produces growth promoting substances such as IAA, GA *etc.* Some Azospirillum species produce siderophores and are able to solubilize phosphorus also. In Azospirillum also crop specific strains are available. (AZ 204 for paddy and AZ SP7 for other crops). Based on the field study conducted over the past 20 years on field inoculation it can be concluded that these bacteria are capable of promoting the yield of agriculturally important crops upto 30 per cent in different soil and climatic regions.

Azolla

Azolla is a floating water fern, which is in symbiotic association with the cyanobacterium *Anabaena azollae* and contributes substantial amount of biological nitrogen fixation to the rice crop. Azolla is used as biofertilizer only for rice crop and it could contribute20-40 kg N ha^{-1} per crop. Azolla can be applied to rice in two ways *viz.*, as green manure @ 5 ton/ha and grown as dual crop along with paddy @ 500kg/ha on 7 Days after planting. Since it is grown along with paddy, it doesn't require extra water or nutrient for its growth and multiplication. Azolla biomass produced in rice fields (approximately 15-20 ton/ha) should be incorporated in the soil on 30th and 45th Day. Azolla play a significant role in soil organic matter buildup also.

Gluconacetobacter

Gluconacetobacter diazotrophicus is an endophytic nitrogen fixing diazotroph occur in the root, stem and leaf tissues of sugarcane, rhizosphere soil and even in cane juice. This biofertilzer can be applied in the form of sett treatment (2kg/

Table 13.1: Biofertilizers and their Applications to different Crops

Sl.No.	*Name of the crop*	*Type of planting*	*Season*	*Biofertilizer recommended*	*Method of application*	*Dose*	*Time of application*
1	Paddy	Transplanted	Kharif	Blue green algae	Dried Soil based algal flakes along with 500g of SSP	10 kg/ha	7-10 days after transplanting (DAT)
		Transplanted	Rabi	Azolla	As dual crop along with paddy	500kg/ha	7-10 DAT allow it to grow along with paddy and incorporate in the soil during I and 2 weeding
		Direct sown	All reasons	*Azospirillum*, PPFM and Phosphobacteria	1. Seed treatment (mix each 200g of biofertilizers with 250 ml rice gruel and mix the seed in the slurry, shade dry)	1kg each/ha	Before sowing
					2. Soil application (mix the biofertilizers with 25 kg FYM/sand and broadcast)	2kg each/ha	Before sowing/ transplanting.
		Transplanted	Kharif and Rabi	*Azospirillum*, PPFM and Phosphobacteria	1. Seed treatment (as mentioned earlier)	1 kg each/ha	Before sowing
					2. Seedling root dipping (mix the biofertilizers in 100 lit of water, remove the seedlings from nursery and dip the roots of the seedlings in biofertilizer slurry for about 20-30 min)	1kg each/ha	Before transplanting
					3. Soil application (as mentioned earlier)	2kg each/ha	Before transplanting

Sl.No.	Name of the crop	Type of planting	Season	Biofertilizer recommended	Method of application	Dose	Time of application
	Paddy- SRI	Nursery	Kharif and Rabi	*Azospirillum*, Azotobacter, PPFM and Phosphobacteria	1. Seed treatment (as mentioned earlier)	1 kg each/ha	Before sowing
				VAM	2. Soil application (as mentioned earlier)	2kg/100m^2	During nursery bed preparation
		Transplanting	Kharif and Rabi	*Azospirillum*, *Azotobacter*, PPFM and Phosphobacteria	1. Seedling dipping (as mentioned earlier)	2kg each/ha	Before transplanting
				Azospirillum, *Azotobacter*, PPFM and Phosphobacteria	2. Soil application (as mentioned earlier)	2kg each/ha	Before sowing or transplanting
2	Other non-leguminous crop (cereals, millets, oilseeds, cotton, mulberry, forage crops	Direct sown	Kharif and Rabi	*Azospirillum*, Azotobacter, PPFM and Phosphobacteria	1. Seed treatment (as mentioned earlier)	1 kg each/ha	Before sowing
				Azospirillum, Azotobacter, PPFM and Phosphobacteria	2. Soil application	2 kg each/ha	Before sowing
3	Pulses	Direct sown	Kharif and Rabi	*Rhizobium*, PPFM and Phosphobacteria	1. Seed treatment	600g each/ha	Before sowing
				Rhizobium, PPFM and Phosphobacteria	2. Soil application	1kg each/ha	Before sowing
						2kg each/ha	

Sl.No.	*Name of the crop*	*Type of planting*	*Season*	*Biofertilizer recommended*	*Method of application*	*Dose*	*Time of application*
4	Sugarcane	Direct sown	Annual	*Azospirillum*, *Azotobacter*, Gluconacetobacter Phosphobacteria.	1. Sett treatment (mix the biofertilizers with water and dip the setts for 30 min and plant the setts)	2 kg each/ha	Before planting
				Azospirillum, Azotobacter, Gluconacetobacter Phosphobacteria	2. Soil application	2 kg each/ha	Before planting
5	Sugarcane	Precision farming	Annual	*Azospirillum*/ *Azotobacter*/ Gluconacetobacter and Phosphobacteria	Biofertigation (200 ml in 200 lit of water)	1 per cent	at 15,30 and 45 DAS.
6	Tree crops	Planting	Annual/ Perennial	*Azospirillum*/ *Azotobacter*/and Phosphobacteria and VAM	Soil application (mix with FYM/ sand) near to the root zone and cover it with soil.	5-10g/pit	Before planting
7	Tree seedlings	Sowing	All seasons	*Azospirillum*/ *Azotobacter* and Phosphobacteria	1. Mixing with the pot mixture and filling in the polybag	5 kg each/ 1000 kg pot mixture	Along with manure application, during pot mixture preparation.
				VAM	2. Mixing with the pot mixture and filling in the polybag	10 kg each/ 1000 kg pot mixture	
8	Grown up trees		All seasons	*Azospirillum*/ *Azotobacter* and Phosphobacteria	1. Spot application (as mentioned earlier)	20-50 g each/ tree	Along with manure application
				VAM	2. Spot application	200 g/tree	

Sl.No.	Name of the crop	Type of planting	Season	Biofertilizer recommended	Method of application	Dose	Time of application
9	Horticulture crops	Fruits	Annual	*Azospirillum*/	1. Spot application	20g/plant	At the time of planting and after 5th month of planting.
				Azotobacter and Phosphobacteria and	2. Spot application	50g/pit	At the time of planting
				VAM			
10	Vegetables and flower crops	Direct sown	Kharif and Rabi	*Azospirillum*, PPFM and Phosphobacteria	1. Seed treatment	1 kg each/ha	Before sowing
					2. Soil application	2kg each/ha	Before sowing
		Transplanted	Kharif and Rabi	*Azospirillum*, PPFM and Phosphobacteria	Seed treatment	1kg each/ha	Before transplanting
			Nursery	VAM	Soil application	100 g/m^2	During nursery bed preparation
			Main field	*Azospirillum*, PPFM and Phosphobacteria	Soil application	2kg each/ha	Before transplanting

ha), soil application (Gluconacetobacter 2 kg/ha) and fertigation methods. If the biofertilizer is applied through drip irrigation method means liquid inoculum should be applied @ 1 per cent at monthly intervals. It can fix nitrogen on an average of 200 kg N ha^{-1}In addition to nitrogen fixation this organism also produce indole -3 acetic acid which serve as a growth promoter.

AM Fungi

Arbuscular Mycorrhiza is a fungal biofertilizer commonly used for nursery (mostly vegetables) crops and trees. Application of AM fungi results in effective utilization of phosphorus, nitrogen sulphur, zinc and calcium from soil and protect the crop from soil borne pathogens. Since it is an obligate parasite this fungi cannot be cultured under laboratory conditions.

Chaper 14

Potential Applications of *Spirulina* in Food and Agro-based Industries

G. Gayathry

Assistant Professor (Agrl. Microbiology),
AC and RI, Kudumiyanmalai, Tamil Nadu

Arthrospira (*Spirulina*) *platensis* is a filamentous cyanobacterium comprising at least 38 species. It belongs to the family oscillatoriaceae, phylum cyanophyta. *Spirulina* exists in saline and aqueous habitats and has been consumed in Africa and Mexico since ancient times (Raja *et al.*, 2008). It is produced by large factories in countries such as China, India and The United States, for the presence of numerous intracellular compounds of great commercial interest, including high contents of proteins, fatty acids (gamma linolenic acid), bioactive polysaccharides and pigments (chlorophyll a, beta-carotene and phycocyanin) (Pulz and Gross, 2004).

Spirulina platensis or its extract show therapeutic properties, such as the ability to prevent cancers, decrease blood cholesterol level, reduce nephrotoxicity of pharmaceuticals and toxic metals and provide protection against the harmful effect of radiation. Besides the large consumption of *Spirulina* for humans, it is also often used as animal feed. Supplementation of feed with *Spirulina* has been trialled in sheep, cattle, swine and poultry (Bezerra *et al.*, 2010). Ruminants have so far been identified as well suited to *Spirulina* supplementation due to their capacity to digest unprocessed algal material (Gouveia *et al.*, 2008). Furthermore, *Spirulina* supplementation has been associated with heightened rumen microbial crude protein production (Panjaitan *et al.*, 2010). It has been hailed as the ''Food of the future'', besides being considered as an ideal food for astronauts by NASA. It represents the second most important commercial microalga for the production of biomass as health food and animal feed (after Chlorella).

Production of *Spirulina*

Spirulina Production would Offer the following Unique Advantages and Possibilities

The technology for mass cultivation and harvest of *Spirulina* is well established. It has undergone two decades of toxicity testing in addition to its known centuries of human use. Microbiological and other safety standards have been established for *Spirulina* products. The two most commonly grown species *Spirulina* (*Arthrospira*) *platensis* and *Spirulina* (*Arthrospira*) *maxima* are free from cyanobacterial toxins and can be grown (under controlled conditions) free of contaminant cyanobacteria by virtue of their adaptation to a highly alkaline environment. Indeed, it is this same property that makes it possible to grow *Spirulina* in land unsuitable for conventional agriculture. Further *Spirulina* has already been popularized through two decades of commercialization as a food and dietary supplement.

Factors Affecting *Spirulina* Production

The large-scale production of *Spirulina* biomass depends on many factors, the most important of which are nutrient availability, temperature and light. These factors can influence the growth of *Spirulina* and the composition of the biomass produced by changes in metabolism.

Nutrients for *Spirulina* Production

Culture media that are used for *Spirulina* can be broadly grouped into three major categories, namely complete synthetic media, those that are based on natural waters enriched by mineral supplementation and liquid wastes such as effluents from fermented wastes or from waste water treatment plants, industries, etc (Vonshak and Tomaselli, 2000). *Spirulina* production as well as its photosynthetic activity and growth-physiology is greatly restricted by the culture media. Two commonly used media under laboratory conditions includes Zarrouk (Z9) and Provasoli's enriched seawater (PES) media. *Spirulina* is largely studied in a bench scale using alkaline culture media namely Z9 medium with high levels of carbonate and bicarbonate that are used as carbon sources and whose correct proportions can provide the desired pH value (about 9.5). As the cells multiply, bicarbonate concentration decreases and leads, consequently, to an increase in pH, which has been successfully controlled by supplying CO_2 to maintain the pH at an optimal range for the *Spirulina* production (Soletto *et al.*, 2008). *Spirulina* media shift cultures were reported by Chen, (2011). Zarrouk (Z9) and Provasoli's enriched seawater (PES) media were re-modified into five different media by adding varied concentrations of NaCl, $NaHCO_3$ and seawater. *S. platensis* grew well in all of these media. The Z9 medium provided the best growth rate at the end of culture and the PES medium provided the best initial growth curve slope. After the media were shifted from PES to Z9, the long algal filaments fragmented to form many short filaments within 5 days of culture. Those short filaments then straightened into less spiral forms due to the ingrowths of active cell division after seven days of culture. A combination of Zarrouk and Provasoli media produced a new medium for ultrahigh-density culture of *Spirulina platensis*. These media shift cultures were easy to operate and ready to use with the lowest cost and high biomass in comparison with other culture methods. Qasim *et al.* (2012) studied the exploration of indigenous *Spirulina* in Sargodha, Lahore and Faisalabad districts of Punjab and its mass cultivation by optimizing the physico-chemical growth requirements. They have concluded that maximum growth of indigenous *Spirulina* was obtained at 30°C and at 1500 lux light intensity. Nitrogen concentrations of 0.625, 1.25 and 1.875 g/L had no effect on the growth, while phosphate concentrations of 0.5, 1.0 and 1.5 g/L had a minimal and gradual effect on growth as the concentrations were increased. Lee *et al.* (2012) studied the two stage culture method for optimised polysaccharide production. The first stage was under 96 μ mol photons m^{-2} S^{-1} light intensity at 28°C and the second stage at 192 μ mol photons m^{-2} S^{-1} at 38°C for 3 days. The first stage increased biomass production and the second stage increased polysacharride production.

Other Mineral Supplements

Phosphorus is a major nutrient required for normal growth of all algae and is essential for almost all cellular processes, *i.e.* the biosynthesis of nucleic acids and energy transfer. The phosphorus concentration is often growth limiting in the natural aqueous habitat, where it occurs as orthophosphate and in organic combinations. The major form in which algae require phosphorus is as inorganic phosphate either as $H_2PO_4^-$ or HPO_3^{2-}. Like phosphorus, sulphur is also vital to all cells because it is a constituent of the essential amino acids, methionine, cysteine, cystine, vitamins and sulpholipids. It is generally provided as inorganic sulphate in the culture medium (Baldia *et al.*, 1991). Makou, (2012) studied the biomass composition of *S.platensis* under various amounts of limited phosphorous condition and observed that the compounds of biomass got altered gradually as the phosphorous became limited. Carbohydrates and lipids increased from 9 per cent to 65 per cent and from 4.9 per cent to 7.5 per cent respectively. The proteins decreased from 46.5 per cent to 25 per cent.

Effect of Temperature on the Growth of *Spirulina*

Temperature is the most important climatic factor influencing the rate of growth of *Spirulina*. Below 20°C, growth is practically nil, but *Spirulina* does not die. The optimum temperature for growth is 35°C, but above 38°C *Spirulina* is in danger. In addition, *Spirulina* thrives in very warm waters of 32°C to 45°C and has even survived in temperatures of 60°C. Certain desert related species will survive when their pond habitats evaporate in the intense sun, drying to a dormant state on rocks as hot as 70°C. In this dormant condition, the naturally blue-green alga turns a frosted white and develops a sweet flavour (Anon, 2005).

Spirulina cells encounter temperature fluctuations, associated with outdoor mass cultivation, that have a relevant effect on biomass yield and the biochemical content of the cells. Some components with pharmaceutical benefits, such as unsaturated fatty acids in membrane lipids, have been shown to play vital roles in the response to temperature change. Physiological changes to the fatty acid content of the cytoplasmic membrane are also observed in *Spirulina platensis*. In comparison with cells maintained at the ideal growth temperature (35°C), *S. platensis* cells synthesize up to 23 per cent more dakua-linolenic acid (GLA; C18:3Δ9,12,6) after a temperature downshift (22°C), whereas the GLA level decreases approximately 30 per cent upon a temperature increase (40°C) (Hongsthong *et al.*, 2003). High-throughput approaches such as the proteomic analyses of the low and high temperature responses in *Spirulina* were performed using two-dimensional differential gel electrophoresis (2D-DIGE) coupled with protein identification by mass spectrometry. Result of the analysis of differential expression has showed that four groups of proteins were found to be in common between *Spirulina* cells under both temperature stress conditions: (i) signal transduction proteins, (ii) chaperones, (iii) stress related proteins and (iv) proteins related to DNA modification and repair. Moreover, two groups of proteins, classified as channeling and secretion and photosynthesis, were detected as differentially expressed proteins only under low-temperature stress.

Light Intensity

Growth only takes place in light (photosynthesis), but illumination of 24 hours a day is not recommended. During dark periods, chemical reactions take place within *Spirulina*, like synthesis of proteins and respiration. Phycobilisomes (PBSs) play the most important role of light harvesting in photosynthesis, that is able to adapt its photosynthetic apparatus not only to light intensity, but also to spectral composition, temperature and availability of CO_2 or other essential nutrients (Li *et al.*, 2003).

Vonshak, (1997) has indicated that light intensity should be provided with care to cultivations, since excess light leads to photoinhibition, that is, a reduction in the photosynthesis rate when the cells exceed the light saturation level. Bezerra *et al.* (2008) has indicated that this phenomenon occurs usually at high photosynthetic photon flux densities (PPFD), but it can take place even at lower values under suboptimal conditions, for example, at temperature lower than the optimum one for growth. It is interesting to note that the lower the PPFD the higher the PE, that is, photosynthetic microorganisms convert light with higher efficiencies. On the other hand, low PPFD values usually lead to light limited cultivations, characterized by low values of cell concentration.

pH

Spirulina is an obligate photoautotroph and has a pH growth range from 8.3 to 11.0. A high alkalinity is mandatory for its growth. *Spirulina* requires a high pH, salinity and bicarbonate concentration. Pelizer *et al.* (2002) has indicated that *Spirulina* can tolerate approximately 7.0 g of NaCl and 50 g $NaHCO_3$ per litre without any measurable ill effects. *Spirulina* can readily tolerate progressive changes in pH. However, it could quickly deteriorate when the pH is changed abruptly, as may happen in a growth medium which is not well buffered. Buffering can be provided with the use of 0.2 M $NaHCO_3$. The pH of the medium further decides the solubility of the carbon dioxide and minerals in the medium.

Spirulina Cultivation Systems

Two important systems are normally used for *Spirulina* cultivation. They are Photobioreactors and open pond systems.

Photobioreactors

Photobioreactors have been successfully used for producing large quantities of *Spirulina* biomass. Two most widely used photobioreactors are open type and closed type. In the case of commonly available open type photobioreactors due to the relatively high depth, the light has to go through thick layers to reach the inner cells (Soletto *et al.*, 2005).Tubular photobioreactors is a type of closed photobioreactor and are usually made of thin transparent tubes which have high area/volume ratios. It consists of an array of straight transparent tubes that are usually made of plastic or glass. This tubular array, or the solar collector, is where the sunlight is captured. The solar collector tubes are generally 0.1 m or less in diameter. Tube diameter is limited because light does not penetrate too deeply in the dense culture broth that is necessary for ensuring a high biomass productivity of the photobioreactor. Microalgal broth is circulated from a reservoir (*i.e.* the degassing column in to the solar collector and back to the reservoir. The solar collector is oriented to maximize sunlight capture. In a typical arrangement, the solar tubes are placed parallel to each other and flat above the ground. Horizontal, parallel straight tubes are sometimes arranged like a fence in attempts to increase the number of tubes that can be accommodated in a given area. The tubes are always oriented North-South. The ground beneath the solar collector is often painted white, or covered with white sheets of plastic, to increase reflectance or albedo. A high albedo increases the total light received by the tubes. Light is better captured by cells in tubular photobioreactors when compared to open type (Ferreira *et al.*, 2010). Instead of being laid horizontally on the ground, the tubes may be made of flexible plastic and coiled around a supporting frame to form a helical coil tubular photobioreactors. Photobioreactors such as the one are potentially useful for growing a small volume of microalgal broth, for example, for inoculating the larger tubular photobioreactors that are needed for producing biodiesel. Other variants of tubular photobioreactors exist, but are not widely used. Pulz (2001) has explained about artificial illumination of tubular photobioreactors is technically feasible but expensive compared with natural illumination. Nonetheless, artificial illumination has been used in large-scale biomass production particularly for high-value products. photobioreactors require cooling during daylight hours. Furthermore, temperature control at night is also useful. For example, the nightly loss of biomass due to respiration can be reduced by lowering the temperature at night. Outdoor tubular photobioreactors are effectively and inexpensively cooled using heat exchangers. A heat exchange coil may be located in the degassing column. Alternatively, heat exchangers may be placed in the tubular loop. Evaporative cooling by water sprayed on tubes, can also be used and has proven successful in dry climates. Large tubular photobioreactors have been placed within temperature controlled greenhouses, but doing so is prohibitively expensive for producing biodiesel. Although open ponds are mainly used for *A. platensis* commercial production, tubular photobioreactors have been widely studied, not only because of their high cell productivity, but also of many other advantages, such as low levels of contamination and better CO_2 solubilization in the media.

Open Pond System

The most widely used system for large-scale outdoor microalgae production is based almost exclusively on open ponds, due to their low cost and ease of operation. Becker (1994) has indicated that open ponds have a variety of shapes and sizes but the most commonly used design is the raceway pond. An area is divided into a rectangular grid, with each rectangle containing a channel in the shape of an oval; a paddle wheel is used to drive water flow continuously around the circuit. They usually operate at water depths of 15-20 cm, as at these depths biomass concentrations of 1.0 g dry weight per litre and productivities of 60-100 mg L^{-1} day^{-1} (*i.e.* 10-25 g m^{-2} day^{-1}) are possible. However, such productivities are not the rule and cannot be maintained on an annual average. Similar in design are circular ponds which are commonly found in Asia and the Ukraine. Algae ponds used in wastewater treatment plants are built in whatever shape best suits the location; these are usually not mechanically mixed but driven by gravity flow. One of the largest of this type of algal pond is Melbourne's Werribee wastewater treatment plant spanning 11,000 ha. In the case of algal wastewater treatment ponds, retaining walls or dug trenches form the basis of the ponds.

Bao *et al.* (2012) studied the feasibility of using an *in situ* carbon supply system for CO_2 addition during large-scale *S. platensis* cultivation in an open raceway pond was studied. The results of the simulation experiments without microalgae indicated that the CO_2 absorptivity in open race way ponds (>84 per cent) was improved by *in-situ* carbon-supply system. The improvement in the average CO_2 absorptivity (to 86.16 per cent and the average CO_2 utilization (to 79.18 per cent) in real cultivations of *S. platensis* confirmed the advantages of this system. It was further concluded that the production cost can be reduced to 40 per cent by using the *in situ* carbon supply system.

Harvesting and Processing

Spirulina harvesting, processing and packing has eight principle stages. The first and foremost is filtration and cleaning. A nylon filter at the entrance of the water pond is needed. In all commercial production processes, filtration devices used for harvesting are basically two types of screens: inclining or vibrating. Inclined screens are 380-500 mesh with a filtration area of 2-4 m^2/unit and are capable of harvesting 10-18 m^3 of *Spirulina* culture per h. Biomass removal efficiency is high, up to 95 per cent and two consecutive units are used for harvesting up to 20 m^3/hour from which slurry (8-10 per cent of biomass) is produced. Vibrating screens can be arranged in double or triple decks of screens up to 183 cm in diameter. Vibrating screens filter the same volume per unit time as the inclining screens, but require only one-third of the area. Their harvesting efficiencies are often very high. At one commercial site a combination of an inclining filter and a vibrating screen has been used. In the process of pumping the algal culture to be filtrated, the filaments of *Spirulina* may become physically damaged, and repeated harvesting leads to an increasingly enriched culture with unicellular microalgae or short filaments of *Spirulina* that pass through the screen readily (Vonshak and Tomaselli, 2000). The slurry obtained after filtration is further concentrated by filtration using vacuum tables or vacuum belts, depending on the production capacity. Secondly, Pre-concentration to obtain algal biomass which is washed to reduce salts content followed by concentration to remove the highest possible amount of interstitial water. Next is the most important neutralization step performed to neutralize the biomass with the addition of acid solution. Then the biomass is disintegrated to break down trichomes by a grinder. Finally dehydration is done by sun drying or spray-drying. This operation has great economic importance since it involves about 20-30 per cent of the production cost. Finally packing and storage. It is usually in sealed plastic bags to avoid hygroscopic action on the dry *Spirulina* and Stored in fresh, dry, unlit, pest-free and hygienic storerooms to prevent *Spirulina* pigments from deteriorating.

Applications of *Spirulina*

Richmond, (1988) has indicated that *Spirulina* is being developed as "food for the future? because of its remarkable ability to synthesise high-quality concentrated food more efficiently than any other alga. Most notably *Spirulina* is 65 to 71 per cent complete protein, with all essential amino acids in perfect balance. In comparison beef is only 22 per cent protein. *Spirulina* provides a high concentration of many other nutrients, chelated minerals, pigments, rhamnose sugars (complex natural plant sugars) trace elements and enzymes in an easily assimilable form. The nutritive value of *Spirulina* is amplified in that it has a relatively low percentage of nucleic acids (4 per cent) as compared with the high percentage of nucleic acid content in bacteria. It is high in Vitamin B_{12}, the muco-protein cell wall are easily digestible unlike the cellulose cell wall found in other nutritional algae, it is non-toxic and has lipids that are made up of unsaturated fatty acid that do not form cholesterol. This makes *Spirulina* a potential treatment for people suffering from coronary illness and obesity. Bhaskar *et al.* (2005) has indicated that *Spirulina* is available as pills, capsules, pastries, blocks and chocolate bars *etc.*, Other *Spirulina* products are formulated for weight loss and as an aid for quitting drug-addictions. The use of *Spirulina's* pigments as colourants has been explored by the pharmaceutical and food industries. The blue pigment, phycocyanin is used by the Japanese as a food colourant. The trend is to substitute artificial colourants with natural ones and since microalgae are the highest source of chlorophyll, the idea is to exploit chlorophyll for this purpose.

Amino Acids and Vitamins from *Spirulina*

The amino acid spectrum of *Spirulina* protein is similar to that of other microorganisms, and when compared to proteins of eggs or milk, it is deficient in methionine, cysteine and lysine Although the amino acid content of *Spirulina* is generally well balanced, it is low in the sulphur containing amino acids and in tryptophan. RNA has been reported to represent 2.2 to 3.5 per cent of the dry weight, whereas DNA represents 0.6 to 1.0 per cent. The total nucleic acid content is therefore 4 per cent of the dry weight (Ciferri, 1983). *Spirulina* contains β-carotene an analogue of vitamin A and possess the capacity to produce two molecules of vitamin A. Being the richest source of vitamin B_{12} Panmol company (Austria) has exploited and commercialised vitamin B_{12} from *Spirulina maxima*.

Pigments from *Spirulina*

The use of *Spirulina* pigments as colourants has been explored by the pharmaceutical and food industries. In *Spirulina* cells, carotenoids, chlorophyll, and phycocyanin (PC) are the three major pigments. Pigments found in *Spirulina* belong to three classes. Chlorophyll *a* comprises 1.7 per cent of the organic cell weight. Carotenoids and xanothophylls comprise 0.5 per cent of the organic weight and the two phycobiliproteins: c-phycocyanin and allophycocyanin normally comprise about 29 per cent of the cellular protein and are quantitatively the dominant pigments in *Spirulina*. PC is (blue) pigment protein (biliprotein) located in the thylakoid system or photosynthetic lamellas in the cytoplasmic membrane. When the cell envelope is broken, thylakoid membrane together with PC are released. PC used in food colourant, nutraceutical and immuno diagnostic applications is mainly extracted from *Spirulina* (Pelizer *et al.*, 2003). Raghu. (2005) has studied the extraction of pharmaceutical grade phycocyanin pigment using Ammonium sulphate precipitation method and the purity per cent was recorded to be 4.81. The molecular weight of the purified compound was found to be 20.1 K Da. The genes responsible for coding the phycocyanin production has been identified as pcyA, cpc E, cpc F and cpc BA. Duangsee *et al.* (2009) carried out the phycocyanin extraction from *Spirulina platensis* using various extraction methods and the effect of pH and temperature on phycocyanin stability. Three extraction methods: sonication, repeated freezing and thawing (RFT) and enzymolysis were carried out on two strains of *S. platensis*, IFRPD1183 (Sp1183) and IFRPD1213 (Sp1213).The results of their study has indicated that sonication was more effective in breaking the cell envelope compared with RFT. The enzymolysis method of phycocyanin carried out using lysozyme under various enzyme concentrations, extraction time and temperature showed that the percentage of extraction efficiency (per cent EE) was significantly affected by extraction time and temperature, but not enzyme concentration. Further extraction of phycocyanin (PC) showed that temperature greatly contributed to EE. Without lysozyme, the maximum EE was obtained at 44°C. PC exhibited the highest stability at pH 5.0. At pH 3, PC was very sensitive to heat and underwent rapid changes.

Lipids and Fatty Acids from *Spirulina*

Pratoomyot, (2005) has indicated that the dominant lipids in *Spirulina* are mono, di- and higher galactosyl-diglycerides and phosphatidylglycerol. The dominant components in the non-polar lipid fraction were free fatty acids. The lipopolysaccharides comprised 1.8 per cent of the dry weight. The carbohydrate content of the lipolysaccharide consisted of hexose, heptose and glucosamine. Palmitic acid is the most abundant fatty acid followed by linolenic and linoleic acids. The carbohydrate content of *Spirulina* was about 15 per cent of the dry weight and hydrolysis yields glucose, sucrose, levulose, glycerol and several polyols. Jublie *et al.* (2012) has isolated stearic acid, a saturated fatty acid from *S.platensis*. The synthesised fatty acid was tested for antidepressant activity in Swiss Albino mice. The stearic acid 1,2,4- thiadiazole showed significant antidepressant activity. Further *Spirulina* is being exploited to produce ^{13}C incorporated Di-hydroxy Acetone (DHA) targeted for the production of instant foods with DHA (Chakdar *et al.*, 2012).

Antioxidant Effects of *Spirulina*

Spirulina is well-known to have antioxidant properties, which are attributed to molecules such as phycocyanin, β-carotene, tocopherol, dakua-linolenic acid and phenolic compounds. Gargouri *et al.* (2012) evaluated the toxic effects of lead acetate on brain tissue and potent protective effects of *S. platensis* in mice experiment. *S. platensis* fed mice showed reduction in lead induced oxidative stress and related damages. It showed protective effects against oxidative stress induced by lead acetate in the liver and kidney of rats. Feeding of *S.platensis* also reduced hepatoxicity induced by cadmium in rats and the effect is suggested to be mediated through its antioxidant properties. *Spirulina* also showed protective effects against nephrotoxicity due to oxidative damage induced by gentamicin.

Antimicrobial Effects of *Spirulina*

The search for cyanobacteria with antimicrobial activity has gained importance in recent years due to growing worldwide concern about alarming increase in the rate of infection by antibiotic-resistant micro-organisms. Various active substances with antibacterial, antiviral, fungicide, enzyme inhibiting, immunosuppressive and cytotoxic and algicide activity have been isolated from cyanobacterial biomass. *S. platensis* produce a diverse range of bioactive molecules, making them a rich source of different types of medicines. The antimicrobial activity of methanolic extract of *S. platensis* was also explained by Demule *et al.* (1996) due to the presence of dakua-Linolenic acid. Compounds such as 1-Octadecene, 1-Heptadeceane found in plants possessing anticancer, antioxidant and antimicrobial activity was also present in acetone extract of *S.platensis.* Babu *et al.* (2002) has studied about the efficacy of *Spirulina platensis* on human pathogens, mosquito and uzifly and has indicated that growth of *Bacillus subtilis, E. coli and Pseudomonas pyocyaneous* was inhibited by methanolic extracts of *S. platensis* and 80-100 per cent extract showed complete mortality of mosquitoes and uzifly. Babu *et al.* (2005) has elucidated about an antiviral principle in *S. platensis* against *Bombyx mori* Polyhedrosis Virus (BmNPV). It inhibits the viral replication in both the cellular and humoral arms of the immune system. They have indicated that silkworm fed with *S.platensis* at 10 per cent concentration showed 90 per cent resistance to BmNPV. Seyoum (1998) has explored that aqueous extract of *S. platensis* partially inhibited HIV-1 replication in human T-cell lines, peripheral mononuclear cells and langerhans cells. Marcel (2011) has illustrated that *Spirulina* supplementation in HIV infected patients play a beneficial role in improving on HIV/HAART associated insulin resistance. WHO projects that *Spirulina* will become a superfood with one of the most curative and prophylactic components of nutrition in 21st century. Al-wathman *et al.* (2012) investigated on the potential antibacterial activities of *S.platensis* and confirmed that the solvent extract of *S.platensis* showed strong antibacterial activity against *Staphylococcus sonnei, S. aureus* and *Bacillus subtilis.*

Spirulina in Biodiesel Production

Biodiesel is developing into one of the most important near-market biofuels as virtually all industrial vehicles used for farming, transport and trade are diesel based. In the past decade, the biodiesel industry has seen massive growth globally, more than doubling in production every 2 years (Oilworld, 2007). The increased demand for vegetable oils for the production of biodiesel has led to significant pressure on the vegetable oil market. Indeed the world's biodiesel industry is currently operating far below capacity due to a lack of feedstock. Second generation microalgal systems are increasingly predicted by international experts and policy makers to play a crucial role in a clean environmentally sustainable future as they have important advantages. Most significantly, these achieve a higher yield per hectare and their ability to be cultivated on non-arable land, thereby reducing the competition with food crops for land.

Feed Stocks of Biodiesel

- ☆ Waste cooking oil, animal fat and oil crops such as corn, soybean, canola, jatropha, coconut and oil palm.
- ☆ In nature, this chemical energy is stored in a diverse range of molecules (*e.g.* lignin, cellulose, starch, oils).
- ☆ Lignocellulose, the principle component of plant biomatter, can be processed into feedstock for ethanol production. This can be achieved by either gasification or by cellulolysis (chemical or biological enzymatic hydrolysis). These processes are currently being developed for second generation biofuel systems (Demain *et al.*, 2005) and are often referred to as 'ligno cellulosic processes'. Similarly, starch (*e.g* from sugarcane) are already being converted into bioethanol by fermentation, while oils (*e.g.* from canola, soy and oil palm) are being used as a feedstock for the production of biodiesel. Microalgae especially Spirulina and Chlorella are able to efficiently produce cellulose, starch and oils in large amounts. In addition, some microalgae and cyanobacteria (which produce glycogen instead of starch) can also produce biohydrogen under anaerobic conditions (Das and Veziroglu, 2001) and their fermentation can also be used to produce methane.

Advantages of Algal Feed Stock

- ☆ Microalgae appear to be the only source of biodiesel that has the potential to completely displace fossil diesel.
- ☆ Unlike other oil crops, microalgae grow extremely rapidly and many are exceedingly rich in oil.
- ☆ Microalgae commonly double their biomass within 24 hrs.
- ☆ Biomass doubling times during exponential growth are commonly as short as 3.5 hrs.
- ☆ Oil content in microalgae can exceed 80 per cent by weight of dry mass.

Conclusion

Spirulina is unusual among algae because it is a "nuclear plant" meaning it is on the developmental cusp between plants and animals. Spirulina is one of the cleanest, most naturally sterile foods found in nature. Its adaptation to heat assures that Spirulina retains its nutritional value even when subject to high temperatures during processing and shelf storage, unlike many plant foods that rapidly deteriorate at high temperatures. The numerous curative and health-promoting properties of Spirulina are truly amazing. This superfood is rich in vitamins, minerals, trace minerals, and antioxidants, all of which make it highly beneficial as an anti-aging, anti-cancer, and super-detoxifying miracle food.With regards to production process, culture media for Spirulina sp. growth and production can make use either of inorganic nutrients ("clean methods") or of a mixture of inorganic and organic residues ("waste methods"), in the form of simple or more complex nitrogen sources. Open pond system of raceway type has been very successful all over the world. The total production of spirulina powder in India is estimated at approximately 300 tonnes and the major producers are Dabur group and EID Parry Ltd. There is a good demand in the domestic as well as international market for both human consumption and also for therapeutic preparations. The world demand for spirulina is estimated at 5000 tonnes per annum and total supply is to the extent of about 2000 tonnes per annum. The world wide sales of spirulina is growing at an astronomical rate of 30 per cent per year. The US leads world production followed by Thailand, India and China. More countries are planning production for its use as a valuable source of protein. Many small and medium scale enterprises has spawned across the country. Still there exists a wide production and application potential for this algal food.

References

Al-Wathman, A., I Ara, R.R. Tahmay and M. A. Tahmir. 2012. Antibacterial extracts of cyanobacterial and green algal isolates of desert soil in Riyadh, Kingdom of Saudi Arabia. Afri. J. Biotechnol., 11: 9223-9229.

Babu, S. M., Gopalaswamy, G and S. Sadasivam. 2002. Efficacy of *S. platensis* on human pathogens, mosquito and uzifly. *In:* Proceedings of national seminar on developments in microbial technology. Dec 12- 13, 2002. GRD Arts and Science College, Coimbatore.

Babu, S. M., Gopalaswamy, G and N. Chandramohan. 2005. Identification of antiviral principle in *Spirulina platensis* against *Bombyx mori* Polyhedrosis Virus (BmNPV). Ind. J. Biotechnol., 4(4): 384-381.

Baldia, S.F., Nishijima, T and Y. Hata.1991. Effect of physico-chemical factors and nutrients on the growth of *Spirulina platensis* isolated from lake Kojima, Japan. Nippon Suosan Gakkaishi, 57: 481-490.

Bao Y., Liu, M., Wu, X., Cong, W and Z. Ning. 2012. In situ Carbon Supplementation in Large-scale Cultivations of *Spirulina platensis* in Open Raceway Pond. Biotechnol Bioprocess Engg., 17: 93-99.

Bezerra, R.P, Matsudo, M.C, Converti, A., Sato, S and J.C.M. Carvalho. 2008. Influence of ammonium Chloride feeding time and light intensity on the cultivation of *Spirulina (Arthrospira) platensis*. Biotechnol Bioeng., 100: 297–305.

Bhaskar, S.U., Gopalaswamy, G and R. A. Raghu. (2005). Simple method for efficient extraction and purification of C-Phycocyanin from *Spirulina platensis* Geitler. Ind J. Exptl. Biol. 43(3): 277-279.

Chakdar, H, H. D. Jadhav, D. W. Dhar and S. Pabbi. 2012. Potential applications of Blue Green Algae. J. Sci. Ind. Res., 71: 13-20.

Chamorro, G., Salazar, M., Araujo, K.G., Dos Santos C.P., Ceballos, G and L.F. Castillo. 2002. Update on the pharmacology of *Spirulina* (*Arthrospira*), an unconventional food. Arch. Latinoam. Nutr., 52(3): 232-40.

Chen, Y. C. 2011. The effect of shifts in medium types on the growth and morphology of *spirulina platensis* (*Arthrospira platensis*). J. Marine Sci. Technol., 19(5): 565-570.

Ciferri, O. 1983. *Spirulina*. The Edible Microorganism. Microbiology Reviews. 47: 551-578.

Demule, M.C.Z, Decaire, G.Z and M.S. Decano.1996. Bioactive substances from *Spirulina platensis.* Int. J. Exp. Biol., 58: 93-96.

Duangsee, R., Phoopat, N and S. Ningsanond. 2009. Phycocyanin extraction from *Spirulina platensis* and extract stability under various pH and temperature. As. J. Food Ag-Ind., 2(04): 819-826.

Ferreira, L.S, Rodrigues, M.S, Converti, A., Sato, S and J. C. M. Carvalho. 2010. A new approach to ammonium sulphate feeding for fed-batch *Arthrospira* (*Spirulina*) *platensis* cultivation in tubular photobioreactor. Biotechnol Prog., 26: 1271–1277.

Ferreira, L.S, Rodrigues, M.S, Converti, A., Sato, S and J. C. M. Carvalho. 2012. Kinetic and Growth Parameters of *Arthrospira* (*Spirulina*) *platensis* Cultivated in Tubular photobioreactor Under Different Cell Circulation Systems. Biotechnol. Bioengg., 109: 444–450.

Gargouri, M., F. G. Goubaa, M. B. Magne, C. Magne, X. Dauvergne, R. Ksouri, Y. Krichan, C. Abdelly and A. E. Feki. 2012. *Spirulina* or dandelion enriched diet of mother alleviates lead induced damages in brain and cerebellum of new born rats. Food Chem. Toxicol., 563(7): 2303-2310.

Gouveia, L., Batista, A.P., Sousa, I., Raymundo, A. and N. M. Bandarra. 2008. Microalgae in novel food products. In: 'Food Chemistry Research Developments. K.N. Papadopoulos (Editor) pp. 1-37. Nova Science Publishers: New York, USA.

Henrickson, R. 2009. Earth Food *Spirulina*. Ronore Enterprises, Inc., Maui, Hawaii, USA.p.180.

Hongsthong A, Deshnium P, Paithoonrangsarid K, Cheevadhanarak S and M.Tanticharoen. 2003. Differential responses of the three acyl-lipid desaturases to immediate temperature reduction occurred in two lipid membranes of *Spirulina platensis* strain. C1. J Biosci Bioengg., 96: 519-524.

Hongsthong A, Sirijuntarut M, Prommeenate P, Lertladaluck K, Porkaew K, Cheevadhanarak S, and M. Tanticharoen. 2008. Proteome analysis at the subcellular level of the cyanobacterium *Spirulina platensis* in response to low temperature stress conditions. FEMS Microbiol Lett., 288: 92-101.

Jublie, S., P. R. Ramesh, P. Danabal, R. Kalinga and N. Muruganantham. 2012. Removal of synthetic dyes by biosorption by *S. platensis*. Eu. J. Medical Chem., 43: 24-29.

Kurdrid, K., J. Senacha, M. Sirijuntarut, R. Yutthanasirikul, P. Phuengcharoen, W. Jeamton, S. Roytrakul, S. Cheevadhanarak and A. Hongsthong. 2011. Comparative analysis of the *Spirulina platensis* subcellular proteome in response to low- and high-temperature stresses: uncovering cross-talk of signaling components. Proteome Science. 9: 39.

Lee M.C.L., Y.C. Chen and T.C. Peng. 2012. Two stage culture method for optimisation of polysaccharide production in *Spirulina platensis*. J. Sci Food Agric., 92(7): 1562-1569

Li, D., Xie, J., Zhao, Y and J. Zhao. 2003. Probing connection of PBS wioth the photosystem in intact cells of *Spirulina* platensis by temperature-induced flourescence fluctuation. Biochimica Biophysica Acta, 1557: 35-40.

Marcel, A.K., Ekali, G. L., Eugene, S., Arnold, O. E., Sandrine, E. D., Weid, D. V., Gbaguidi, E., Ngogang, J and J. C. Mbanya. 2011. The effect of *Spirulina platensis versus* soyabean on insulin resistance in HIV-infected patients: A Randomised Pilot Study. Nutrients, 3: 712-724.

Mezzomo N., Saggiorato, A. G., Siebert R, Oliveira P.T, Lago, M C., Hemkemeier, H., Costa, J. A. B., Bertolin, T. E. and L. M. Colla. 2010. Cultivation of microalgae *Spirulina platensis* (*Arthrospira platensis*) from biological treatment of swine wastewater.Ciênc. Tecnol. Aliment. Campinas, 30(1): 173-178.

Natalia, M., A. G. Saggiorato, R. Siebert, P. O. Tatsch, M. C. Lago, M. Hemkemeier, J. A. V. Costa, T. E. Bertolin and L. M. Colla. 2010. Cultivation of microalgae *Spirulina platensis* (*Arthrospira platensis*) from biological treatment of swine wastewater. Ciênc. Tecnol. Aliment Campinas, 30(1): 173-178.

Nejad, S. 2011. Biomass and lipid productivities of marine microalgae isolated from the Persian Gulf and the Qeshm Island. Biomass Bioenerg. 35: 1935-1939.

Panjaitan, T., Quigley, S.P., McLennan, S.R and D.P. Poppi. 2010. Effect of the concentration of *S. platensis* algae in the drinking water on water intake by cattle and the proportion of algae bypassing the rumen. Anim. Prod. Sci., 50: 405-409.

Pelizer, L. H., Carvalho, J.C.M. Sato, S and I. D. O. Moraes. 2002. *Spirulina platensis* growth estimation by pH determination at different cultivations conditions. Electronic J. Biotechnol., 5 (3): 250 – 257.

Pratoomyot, J., Srivilas, P and T. Noiraksar. 2005. Fatty acids composition of 10 micro algal species. The songklanakarin. J Sci Technol 27: 1179-1187.

Pulz, O. 2001. Photobioreactors: production systems for phototropic microorganisms. Appl Microbiol Biotechnol., 57: 287-93.

Pulz, O and W. Gross. 2004. Valuable products from biotechnology of microalgae. Appl Microbiol Biotechnol., 65: 635–648.

Qasim, M., Najeeb, I., Rashee, M., Ali Shahzad K., Ahad A., Fatima, Z and Z. Anwar. 2012. Physico-chemical growth requirements and molecular characterization of indigenous *Spirulina*. Afri. J. of Microbiol Res., 6(11): 2788-2792.

Rachen, D., Natapas, P and S. Ningsanond. 2009. Phycocyanin extraction from *Spirulina platensis* and extract stability under various pH and temperature. As. J. Food Ag-Ind., 2(4): 819-826.

Raghu, R. A. 2005. Ph D Thesis, Tamil Nadu Agricultural University (TNAU), Coimbatore, Tamil Nadu, India.

Raoof, B., Kaushik, B. D. and Prasanna, R. 2006. Formulation of a low-cost medium for mass production of *Spirulina*. Biomass and Bioenergy. 30(6), 537-542.

Raja, R., Hemaiswarya, S., Kumar, N.A., Sridhar, S and R. Rengasamy. 2008. A perspective on the biotechnological potential of microalgae. Crit. Rev. Microbiol. 34: 77-88.

Richmond, A. 1988. Microalgae of economic potential. *In: CRC handbook of Microalgal mass culture*. CRC Press, Inc: Boca Raton, Florida, 199-244.

Seyoum A., Amha, B., Timothy, B. W and Ruth, R. M.1998. Inhibition of HIV-1 Replication by an Aqueous Extract of *Spirulina platensis* (*Arthrospira platensis*).Journal of Acquired Immune Deficiency Syndromes and Human Retrovirology.18(1): 7-12.

Soletto D, Binaghi L, Ferrari L, Lodi A, Carvalho, J.C.M, Zilli, M and A. Converti. 2008. Effects of carbon dioxide feeding rate and light intensity on the fed-batch pulse-feeding cultivation of *Spirulina platensis* in helical photobioreactor. Biochem. Engg. J. 39: 369-375.

Teeranusonti, P., Satirapipathkul, S and P. Duangsri. 2010."Microalgae cultivation using waste water from pickling factor", The 2nd Regional Conference in Bioteachnology; Research and Development on Food Biotechnology, Phnom Penh, Cambodia, February 11-12, 2010.

Vonshak, A. 1997. *Spirulina platensis* (*Arthrospira*): Physiology, Cell Biology and Biotechnology. Taylor and Francis, London, ed, 1997.

Vonshak, A and L.Tomaselli. 2000. *Arthrospira* (*Spirulina*): Systematics and Ecophysiology. *In:* Whitton, B.A., Potts, M. (Eds.), Ecology of Cyanobacteria. Kluwer Academic Publishing, The Netherlands, pp. 505- 523.

Chapter 15

Mulberry and Silkworm: Pharmaceutical Value Perspective

P. Mohanraj, B. Bebitha, C.A. Mahalingam

Tamil Nadu Agricultural University,
Coimbatore – 641 003, Tamil Nadu

Introduction

As health becomes a great concern of people now a day, many natural ways to take care of one's health are gaining more and more popularity. Mulberry and silkworm is having food and medicine and it has been scientifically proved that the values are either equal or better than those made chemically, but have no side effect to our health.

Mulberry Chemical Constituents

Leaves

Flavonoids, steroids, volatile oils, amino acids, vitamins, oxalic acid, citric acid, succinic acid, malic acid, tartaric acid, fumaric acid, palmitic acid, ethylpalmitate

Stem

Steroid sapogenine

Root

Antihelminthetic, Astringent, Deoxnojirimycin

Medicinal Value of Mulberry Plants

Mulberry and its Therapeutic Values

Table 15.1: Vitamin Content in Mulberry Leaf and its Therapeutic Values

Sl.No.	*Name of the Vitamin*	*Content Available in Mulberry (mg)**	*Healing and Therapeutic Properties*
1.	Retinol (Vitamin-A)	-	Eye problems, acne, other skin disorder (Warts, boils, rashes and carbuncles).
2.	Thiamine (Vitamin-Bl)	0.06	Life saving in treatment of cardiovascular disease related to beriberi and infantile beriberi, Heart disease, Gastro intestinal disorders, Intreatment o Alcoholism, Insomnia and Stress.
3.	Riboflavin (Vitamin-62)	0.13-0.21 (mg)	Cataract, Eye ailments, Nervous depression, General debility, Digestive disturbance.
4.	Niacin (Vilamin-B3)	0.60-0.99 (mg)	Provide relief in Pellagra Migraine Headache, High blood cholesterol Arteriosclerosis and Diarrhea, High blood pressure.
5.	Panlothenic acid (Vitamin-B5)	0.16-0.35 (mg)	Arthritis, Infections, Skin Disorders Premature Greying, Stress.
6.	Pyridoxine (Vitamin-B6)	0.43-0.50 (mg)	For treatment of Diabetes Haemorrhoids, Vaginal Bleeding, Stress and Insomnia, Morning Sickness, Travel Sickness, Convulsions in infants and women.
7.	Biotin (Vitamin-B8)	0.02-0.08 (mg)	For Acne and Seborrheic, Eczema, In reducing and controlling excessive hair loss in male Alopecia
8.	Folic acid (Vilamin-B9)		For Magaloblastic Anaemia during pregnency and infants, Recurrent Abortion, Menial Retardation, Gout and Brown Spols on the skin, sprue.
9.	Choline	9.3-15.5 (mg)	For Nephritis, Liver damage, High Blood Pressure, Heart disease.
10.	Inositol	0.40 (mg)	For Premature Greying, Baldness, Heart disease.
11.	Vitamin-C (Ascorbic Acid)	200-300 (mg)	For Common Cold Infections, Stress, Cancer, Athero-sclerosis, Nephritis, hematemesis (vomiting of.blood), Nose-bleeding, Bleeding piles, Melanea (Black stools caused by blood in the intestines), Erythema nodosum (Red nodules in the skin), Anoixia (loss of smell), Deep Corneal ulcers, Perio dontitis (bleeding from the gums), Delayed growth and development, Dental Caries.
12.	Vitamin-D		Rickets, Arthritis, Tooth decay, Bone
13.	Tocopherol (Vitamin-E)		Sterility, Abortion and Anomalies, Dismenorrhoea, Menopause, Old-age, Heart-disease, Varicose Veins, Helps to prevent high cholesterol levels. Improves plalolcls functioning to prevcnlblood clots, breast cancer, fibrocylic disease of breasts, coronary heart disease by strengthening the lining of arteries. Boosts the immune system of the body. Protects against the pollutant air, cataract, retinal regeneration, aithrilis and leg cramps.

Source: Rajendra *et al.,* 1993.

Medicinal Value of Fruits

Morus Fruit Produce Medicinal Effects

The Chinese Materia Medica features numerous materials that are used as foods, including grains (wheat, millet, *etc.*), fruits (jujube, walnut), beans (soybeans, mung beans), and meats (oyster, pork). From the perspective of Western medicine, these are all useful as nutrient sources with obvious value to those suffering from various nutrient deficiencies, but they are usually not considered to have a therapeutic value. On the other hand, it is found by modern research that some foods contain additional non-nutrient ingredients that can have a distinct health impact. A good example is the isolation and characterization of isoflavones from soybeans. Soy is considered an excellent source of protein, and also a source of vitamins (especially vitamin E in its oil); the isoflavones, if consumed in sufficient quantity, can have a significant hormone-like action. Though not studied intensively, mulberry fruits appear to contain one main class of non-nutrient active constituents, which are the anthocyanins. In particular, it is known to contain cyanin (the structure presented here), which contributes the red pigment that gives the fruit a red to purple color. The content in ripe fruits is about 0.2 per cent; an ounce of fruit would provide about 60 mg anthocyanins. The dried fruits are used in doses of 9-15 grams per day in decoction, and this can yield about 90-150 mg of anthocyanins. In Chinese diets, this component may have been low, in which case, such herbal supplements (decoctions or juices) can be an important source.

Health Benefits and Action of Fruits

Deficiencies in Chinese diets may have contributed to those problems mentioned as being remedied by morus fruit, such as anemia, constipation, premature graying of hair, *etc.* While morus fruit contains some nutrients (e.g., small amounts of calcium, iron, vitamin C, and B-vitamins), the anthocyanins may have improved blood circulation and other body functions to alleviate some symptoms that arise under the deficiency conditions.

Medicinal Value of Root

The root possesses antihelminthetic and astringent (stopping bleeding) properties. An alkaloids deoxyjirimycin (DNJ) is extracted form the root of black mulberry and this found to inhibit the enzyme glycosidase and it is considered to be a potential medicine against AIDS.

Commercial Product

Natural Organic Herbal Tea

Mulberry tea comes to us from the heart of South East Asia. Grown in the mountains of Laos, in a city called Vang Vieng. Where Lies a back-to-nature organic farm. The Mulberry plant and it's parts are well known throughout Asia for it's medicinal values. Here is a condensed list of it's health benefits:

- ☆ It contains all 18 amino acids.

- ☆ Source of high calcium and magnesium Vit A, Vit B1, B2, Zinc,Potassium, Vit C.
- ☆ An important antioxidant containing Quercetin, Rutin.
- ☆ A proven weight loss tea.

Paste Chlorophyll

Two types of Chlorophyll "A" and "B" in 3:1 ratio are being extracted from the silkworm faeces. The faeces is treated with alcohol using a stainless steel plant. Method of extraction is simple and establishment expenditure is within the reach of average Chinese sericulturists. The government also provides financial support to such ventures.

Uses

Paste Chlorophyll is being utilized as raw material in chemical industry and to make many kinds of water soluble chlorophyllin. It has a potential market in Japan and every year tones of paste Chlorophyll is exported to Japan from Zhejiang province alone. Reportedly, 40 tonnes of paste Chlorophyll can be extracted from 1000 tonnes of silkworm faeces.

Silk Worm

- ☆ Potential medicinal source of proteins, vitamin B1,B2 and E, diaphause hormone, amino acid
- ☆ A part of antibacterial and antihistaminic preparation
- ☆ Silk worm pupa used-hepatitis, acute pancreatitis, chronic nephritis, stomach and gastric disorder, blood chloestrol

Medicinal Values of Silkworm

Silk Proteins in Biomedical Research

It is well-known that fibroin of silkworm shell is utilized as a textile fibre on account of its high tensile strength, softness and affinity for dyes *etc.* However, on account of good physico-chemical properties and relatively inert immune response, the silk fibroin has also been used as surgical sutures since long. Research reports indicate that silk fibroin possesses properties such as, water vapour permeability, oxygen permeability, blood compatibility, minimal inflammatory reaction *etc.* and suggests its use as one of the possible biomedical and tissue material for enzyme immobilization, cell culture, drug delivery systems, wound dressing. It also acts as a pharmaceutical agent and reportedly lowers the blood glucose levels and enhances alcohol metabolism of the liver. The other biomedical applications of silk fibres, are silk hydroxyapatite composite prosthesis developed for bone regeneration, protective gauges for treatment of skin burns and bioactive textiles with antibacterial activity. Further, silk based devices are reported to assist in repairing of damaged epidermis (artificial skin) or used as scaffold for tissue' regeneration, as biodegradable films and gels for filtration and for drug delivery.

Major Applications

The major applications of silk fibroin, sericin and protein in the biomedical research, are *Silk fibroin as wound dressing material:* It was found that mechanical properties of silk fibroin can be enhanced by blending with other polymers such as poly ethylene glycol (PEG) and semi-interpenetrating polymer networks (SIPNS). The quick wound healing properties of the silk fibroin are attributed to rapid collagen formation induced by it, which is non-cytotoxic to the tissues.

Silk Sericin in Cosmetics

Silk sericin constituting about 25 per cent of the cocoon shell is removed through degumming. As it has got good water retention capacity due to presence of several hydroxyl groups, researchers showed its possible use in cosmetic industry.

Silk sericin polymeric nano particle: Nano particles are colloidal particles below the size of 1 and are utilized in separation techniques, clinical diagnostic assays, drug delivery systems and cosmetics. These particles have advantage over others on account of easy purification, sterilization, drug targeting possibilities and sustained drug release action.

Silk Protein gels for Drug Delivery Systems

The poloxamer hydrogel, a typical polymeric surfactant has been used for a long time as drug delivery system due to their good biocompatibility and less toxicity. Reports indicate that silk fibroin/poloxamer hydrogel composition may be best used as drug delivery vehicles. Study 'conducted with riboflavin as model drug indicated sustained drug release action.

Silk Fibroin for Cell Culture and Wound Healing

Studies using fibroblasts from normal human skin, indicated that fibroin from silk gland and fresh cocoon was reported to accelerate cell cultivation. Further, in an experiment conducted on wound healing on a dog, it was found that amorphous silk fibroin film (ASSF) actually accelerated wound healing as it maintained required moisture environment of the wound. Research is under progress to show silk proteins as functional materials. Hence, sericin and fibroin polypeptides are very important not only for utilization of silk proteins but also as basic research material for their use in biomedical research.

Pupa Skin – A Useful Waste

Chitin

Micro-crystalline Chitin

It finds use as an additive to increase the loaf volume of wheat flour bread from about 4.9 to 6.0 cm^3/gm. It has more water binding capacity than micro-cellulose, which In turn, has slightly more fat binding capacity than micro-crystalline chitin [D. Knorr, 1982].

The use of chitin preparations for the post-operative treatment of the intranasal operations such as, conchotomy, deviatomy, polypectomy *etc.* has showed that it is easier to use, less hemophase, greater pain relief and hastens healing of wounds [K. Kifune, 1992].

Chitosan

It shows potent antimicrobial activities against various bacteria and fungi, Chitopolv is a polynosic fibre blended with chitosan. It has antimicrobial activities against *Staphylococcus aureus, Klebsiella pneumortiae, Aspergillus niger etc.* Chitopoly is widely used as blended spun -yarn and mixed with cotton, polyester, acryl and rayon. Chitosunny, blend of chitopoly and cotton shows antifungal activity against *Trlchophyton equinum.* Therefore, chitosunny finds use in socks for treatment of atheletes foot [Hiroshi *Seo, 1993*).

Low molecular chitosan [LMCS] exhibits inhibiting activity of colonization of Streptococcus mutans on tooth surface of rats and buffering activity against acids produced by plaque bacteria, thereby, inhibiting dental caries increment in rats. LMCS is effective as a food additive to control carcinogenicity of foodstuffs [K. Shibasaki, 1992].

Chitosan in its bioactivity, has many applications (P.A. Sandford 1992), while it performs the functions of immunoadjuvant as an antiviral agent, it stimulates lymphokine, as antiviral and antitumor agents. As a bacterio-static agent, it binds infectious bacteria and essential minerals as well as inhibits growth. On one side. It binds yeasts and molds as fungistatic agent,it stimulates anti-fungal agent in plants, on the other hand. Also, as an anti-sordes agent it prevents adherence/removes carcinogenic bateria from teeth.

Chitin, Chitosan and their Derivatives

These have applications as wound dressings, controlled release of drugs, and contact lenses [P.A. Sandford, 1992].

Chitin and chitosan are useful for enhancement of dissolution properties of poorly soluble drugs such as, griseofulvin, phenytoin, Phenobarbital, predni-solone, flufenamic acid and indomethacin.

China has already started marketing of chitosan based products making head-way in the field. The useful usitlisation of silkworm pupa skin to produce chitin and its derivatives will also lead to the reduction in the cost of reeling and seed production resulting in the low price of finished silk goods.

Proteolytic Enzymes in Cancer Therapy

The clinical research that currently exists on proteolytic enzymes suggests significant benefits in the treatment of many forms of cancer.2 Specifically these studies have shown improvements in the general condition of patients, quality of life, and modest to significant improvements in life expectancy. Studies have consisted of patients with cancers of the breast lung, stomach, head and neck, ovaries, cervix, and colon; and lymphomas and multiple myeloma. These studies involved the use of proteolytic enzymes in conjunction with conventional therapy (surgery, chemotherapy and/or radiation) indicating that proteolytic enzymes can be used safely and effectively with these treatments. Proteolytic enzymes are not recommended for at least two days before or after a surgery as they may increase the risk of bleeding. Proteolytic enzymes have been shown to be quite helpful in speeding up post-surgical recovery and relieving a complication of surgery and radiation known as lymphedema.

Silkworm Commercial Products

Silkworm Amino Acid Powder

Silkworm pupa amino acids are yellowy powder, made from defatted silkworm pupa. It's soluble in water, pH value between 4 to 6.5, and positive ninhydrin reaction.

Applications

1. Silkworm pupa protein is complete protein, containing a variety of amino acids necessary for health, including 8 kinds of amino acids essential to human body. Up to the quality standards of FAO/WHO, its nutritive value is super than that of eggs and milk, rich zinc content reach to 19.025mg/100g
2. Its good solubility in water enables it to produce transparent beverage. It has been widely used in health food industry, such as amino acid drink, granule, capsule, fruity beverage, and kinds of good additives. It is also used as raw material for medicinal amino acid and single amino acid.

Serratiapeptidase Enzyme - Silkworm

Conditions Serrapeptase Help:

Pain of any kind

- ☆ Arthritis Back Problems.
- ☆ Lower Back Problems.
- ☆ Neck Diabetes.
- ☆ Leg Ulcers.
- ☆ Repetitive Strain.

References

Ahmad RS, Sharma SB. Biochemical studies on combined effects of garlic (*Allium sativum* Linn,) and ginger (*Zingiber officinate* Rosc.) in albino rats. Indian JExp Biol. 1997; 35: 841-843.

Anduallu.B and N.Ch.Vardacharyulu.(2001). Effect of mulberry leaves on diabetes, Int. J. Diab. Dev. Countries, Vol 12: 147–151.

Bart C. Hypoglycemic action of the leaves of Morus Alba. Compt. Rend. Soc. Biol 1932; 109, 897–9.

Berg Meyer H U, Bernt E. *In:* Methods of enzymatic analysis, 2nd ed. Berg Meyer, HU., *Ed;* Verlag-Chemie weinheim. Academic press, Inc., New York, 1974, Vol 2. 727-52.

Burman TK. Isolation and hypoglycemic activity of glycoprotein moran A from mulberry leaves. Planta Med 1985; 6: 482-4.

Caspary W F, Rhein A M, Creutzfeldt W. Increase of intestinal brush border hydrolases in mucosa of streptozotocin-diabetic rats. Diabetologia 1972; 8: 412-4.

Chang AY, Schneider DI. Blood glucose and gluconeogenic enzymes. Biochem. Biophys. Acta 1972; 200: 567-8.

Chatterjee G K, Burman T K, Pal SP. Antiinflammatory and antipyretic activities of Morus indica. Planta Med 1983; 48: 116-9.

Data.R.K. (1994).Silkworm to produce human vaccine, *Indian silk*, vol 11, pp 33-35.

Eross J, Kreutzmann D, Jimenez M. *et al.*, Colorimetric measurement of glycosylated protein in whole blood cells, plasma and dried blood. Am Clin Biochem 1984; 21: 519-22.

Gutman AB, Gutman EB. Estimation of acid phosphatase activity of blood serum. J Biol Chem 1940; 136: 201-9.

Hugget A St G, Nixon DA. Use of gluco1se oxidase,peroxidase and o-ianisidine in the determination of blood and urinary glucose. Lancet 1957; 273: 366-70.

Iyer UM., Mani UV. Studies on the effect of curry leaves supplementation (Murraya kenigii) on lipid profile,glycated proteins and amino acids in non-insulin dependant diabetic patients. Plant Foods for HumanNutr.1990; 40: 275-82.

Kamble SM, Kamiakar PL, Vaidya S, Bambole VD.Influence of Coccinia indica on certain enzymes in glycolytic and lipolytic pathway in human diabetes, IndianJ Med Sci 1998; 52(4): 143-6.

Kelkar SM, Bapat VA,Ganapathi TR,Kaklij GS, Rao PS,Heble M R etermination of hypoglycemic activity in Morus indica L. (mulberry) shoot cultures.Current Science 1996; 71(1): 71-2.

Koundinya, P.R and K. Thanavelu. (2005).Silk proteins in biochemical research, Indian *silk vol 17*, pp. 5–6.

Latner A. Clinical Biochemistry 2nd ed. W.B. Saunders., Philadelphia,1958, p. 47.

Murray R K, Granner DK, Mayes PA, Rodwell N Harper's Biochemistry 23rd ed. Prentice Hall of India pvt. Ltd., NewDelhi, India.1993; p. 58.

Oakley WG, Pyke, DA, Taylor KW. Clinical diabetes and its bio-chemical basis. Blackwell Scientific Publications, Oxford 1968, p. 452.

Ramachandran K. The useful plants of India. Publications and Information Directorate, CSIR. Hill side road, New Delhi, 1986, p. 381.

Reid N W. Pharmaceutical aspects of dietary fibre.Dietary fibre: Chemical and Biological aspects. Royal Society of Chemistry, Cambridge,1990, p. 340.

Sailaja, Y.R. Biochemical studies during maturation of reticulocytes to erythrocytes in type 2 Diabetes. Ph.D., thesis, Sri Krishnadevaraya University, Anantapur, 2000.

Sharma RD, Sarkaar A, Hazra. DK. *et al.*, Use of fenugreek seed powder in the management of non-insulin dependent diabetes mellitus. Nutr Res 1996; 16 (8): 1331-9.

Sheela CG, Augusti KT, Antidiabetic effects of S-allylcysteine sulphoxide isolated from garlic Allium sativum Linn.Indian J Exp Bio 1992; 30: 523-6.

Shetty, K.K and B.N. Mohan. (2003). Mulberry green tea for good health, *Indian Silk.* 14: 27–28.

Chapter 16

Role of ICT in Agriculture Extension: Experiences

M. Balarubini[1] and C. Karthikeyan[2]

[1] Research Associate, e-Velanmai Scheme, DoEE,
[2] Professor (Ag. Extension), e-Extension Centre,
Tamil Nadu Agricultural University, Coimbatore

Introduction

The goals of agricultural extension include transferring information from the global knowledge base and from local research to farmers, enabling them to clarify their own goals and possibilities, educating them on how to make better decisions, and stimulating desirable agricultural development (van dan Ban and Hawkins 1996). Thus extension services provide human capital–enhancing inputs, including information flows that can improve rural welfare—an important outcome long recognized in the development dialogue (Feder, Just, and Zilberman 1986; Roberts 1989). That interest continues in contemporary dialogue, as evident in the workshop on public extension services convened by the World Bank, the U.S. Agency for International Development, and the Neuchatel Group to review recent approaches to revitalizing extension services (World Bank 2002). Feder, Willett, and Zijp (2001) identify eight interrelated characteristics of public extension systems that jointly result in deficient performance, low staff morale, and financial stress. These characteristics provide a framework for analyzing the performance of different levels of extension personnel, the system as a whole, and the underpinnings of different organizational forms and for predicting their likely performance.

However, the rural people still lack basic communication infrastructure in accessing crucial information in order to make timely decisions. The application of ICT in agriculture generates possibilities to solve problems of rural people and also to promote the agricultural production by providing scientific information timely and directly to farmers. Here are some benefits of ICT in agriculture.

- ✰ Introduction of mobile phones has brought about a tremendous change in agriculture sector resulting into dramatic improvement in the efficiency and profitability of the agriculture industry. The spread of mobile phone service allow farmer to land their product timely and directly to the market where wholesalers are ready to purchase them without presence of middle man. This situation reduced waste from between 5-8 per cent of total product to close to zero and increased average profitability by around 8 per cent.
- ✰ Radio and television has been another input in communication technology used widely by many farmers, they have been used by farmers, entrepreneurs, extension workers and other stake holders to disseminate information on various innovation in agricultural technology
- ✰ The internet is also an emerging tool with potential to contribute in agriculture sector and in rural development. Internet enables rural communities stay up to date and to receive information about the market and other necessary information in the industry. Internet can facilitate dialogue among communities and help to share information between government planners, development agencies, researchers, and technical experts. The Internet has proven valuable for the development of agriculture in developing countries like Tanzania.

ICTS and Agricultural Extension: What for?

During plenary discussions, participants at the Observatory noted many times that ICTs in the service of improving rural livelihoods cannot be constrained only to agricultural extension. Participants described their objectives for agricultural extension as: having a strong focus on improving the wellbeing of rural communities and rural families; reducing poverty; sustaining environmental resources; and achieving food security. The broader context of improving rural livelihoods was a continuing theme. As Carl Greenridge (2003) noted in his opening address to participants, the important question is how to harness ICTs meaningfully to remedy the many ailments and challenges facing the rural poor.

Clare O'Farrell (2003) highlighted the importance of a focus on rural livelihoods, noting that ICTs can be used to enhance rural livelihoods because they:

- ✰ Can empower rural people by amplifying their voices
- ✰ Are enabling tools that can help rural poor women and men to capitalise on emerging opportunities, especially in education and income generation
- ✰ Can be used to cushion shocks and disasters, such as disease and hunger.

This focus on rural livelihoods over agriculture-specific extension services was also a key theme of the pre-consultation email dialogue (Ballantyne and Bokre, 2003) which defined extension as aiming to:

- ☆ Improve the wellbeing of individuals and communities
- ☆ Improve agriculture and the social, economic and political status of rural communities
- ☆ Improve the wellbeing of farm families
- ☆ Improve productivity and livelihoods for farmers
- ☆ Increase and improve farmers' incomes and productivity on a sustainable basis
- ☆ Enhance farmers' production

ICT for Agriculture Technology Dissemination

The main phases of the agriculture industry are crop cultivation, water management, fertilizer application, fertigation, pest management, harvesting, post harvest handling, transporting of food/food products, packaging, food preservation, food processing/value addition, food quality management, food safety, food storage and food marketing.

All agricultural extension and farmer-outreach programs face three major challenges *viz.* ensuring cost-effective outreach, designing solutions tailored to needs of individual farmers and cultivating an image that is farmer-friendly. Large sections of the farming community, particularly the rural folk, do not have access to the huge knowledge base acquired by agricultural universities, extension centers and businesses. However, internet and mobile networks have the potential to provide agro-information services that are affordable, relevant to needs (timely and customized), searchable and up to date.

Information Communication Technology can provide vital access to information, markets by connecting the rural poor and marginalized to the world's information resources and opportunities. However, not all persons have access to this information. The inequality in opportunities presented by ICT is widest between urban and rural groups, rich and poor, men and women and the educated and uneducated. Despite this, ICT use in rural areas is increasing, such as the internet and cell phones and the individual, community and national benefits they bring by making information available at the fingertips are forever emerging.

Among various initiatives required to improve farm productivity, four critical areas are highly conducive for usage of ICT. IT can help reach the large number of farmers, which otherwise is not possible. The areas where ICT can help are:

- ☆ Farmer Education.
- ☆ Back up services
- ☆ Commercial Information and
- ☆ Help in selling produce

By increasing access to better opportunities to the rural farmer the economy gains in terms of better integrated and competitive product markets, the local family amplifies its opportunity for income; and the society gains from the spillover effects of less poverty and more economically and socially productive citizens. For example, research from a 'Village Pay Phone' project in Bangladesh indicates that the introduction of telephones to the village allowed the villagers to eat well all year round compared to only 9.9 months when there were no phones. Benefits such as these become diffused manifold in the economy and the society.

Today farmers seem to be more innovative and extension agencies/personnel have become laggards. Extension personnel are unable to creatively respond to the change taking place in the environment and remain duplication and tradition bound. Farmers need dynamic information relating to agricultural rural development. There is presently a gap between what farm families need by way generic and dynamic information and what the conventional extension agencies are able to provide. Therefore to satisfy the need of farmers and farming communities, Information and Communication Technology would be very effective.

Farmer Education

India's average farm productivity in food grains is nearly one third of that of a comparable economy like China. However, this average is a result of widely varying levels of productivity across the country. The highest productivity figures achieved with Indian seeds and package of practices in adequately irrigated lands are comparable to those reached anywhere in the world. Thus, the pressing need is to reduce the spread and pull the average up. Extension services of the government have been resource intensive, limited by the extent of actual coverage, competence, motivational and diligence levels of the individuals and their ability to stay updated with latest developments. If we analyse the difference between the farming practices in developed nations like USA and developing country like India, the biggest difference is in availability of information. The gap between the 'information-rich' farmers of Punjab and 'information poor' farmers of Jharkhand can be bridged by IT. A farmer's need can be divided into three hierarchical levels as follows.

Backup Services

It is not sufficient to provide the farmer exhaustive information through web-portals, his specific queries have to be replied through on-line chats, he has to be provided with early warnings regarding pest onset and weather forecasts for his local areas using remote sensing. Once the web portals attract large number of farmers, lot of new set of industries interested in providing services like sowing, transportation and mechanized farm operations like pest control or harvesting will get created.

Once, direct contact of farmers becomes feasible with the help of IT, another extremely and critically important input *i.e.* farm credit sector can be handled much easily.

Commercial Information

Real time information on all the agri-inputs like seeds, and crop protection chemicals, nearest vendors, international sources can be provided. Once the service providers feed farmer with all the information, he can intelligently decide to carry out certain operations on his own or outsource them.

Help in better Price Realization

Most of the small farmers sell their produce to middlemen or in the nearest mandis where the middlemen decide prices. The farmer has virtually no interactions with the *taluka* nor does he know the prices ruling at nearby markets. By making commodity prices and market information on a real time basis available on the Internet, the farming community can be provided with choices that they lack today. This will ensure better price realization and stimulate a drive towards better productivity.

IT enablement can also enormously benefit farmers who grow cash crops by providing forecasted information on future prices of commodities. This will prevent the tendency of farmers to jump into a decision on the basis of ruling price levels and later on discover that the prices have crashed when they are ready to sell their produce at the end of the season. Information on likely future prices of commodities can avert this disaster to a large extent.

Some successful ICT applications in Indian agriculture are as follows:

e-choupal

Indian Tobacco Company (ITC's) International business division is one of India's largest exporters of agricultural commodities has conceived e-choupal has launched in June 2000 as a more efficient supply chain aimed at delivering value to its customers around the world on a sustainable basis. ITC's e-choupal is a unique example of using ITC's for agricultural development; e-choupal has already become the largest initiative among all internet based interventions in rural India. e-choupal link rural farmers directly for the procurement of agricultural/aquaculture produce like soya, coffee, prawns *etc.* eliminating the role of the middleman. The principle of the e-Choupals is to inform, empower and compete. There are 6,500 e-choupals today. ITC Limited is adding 7 new e-Choupals a day and plans to scale up to 20,000 e-choupals by 2012 covering 100,000 villages in 15 states, servicing 15 million farmers.

aAQUA.org

aAQUA-which stands for *almost **All Questions Answered*** is a farmer-expert Q and A database supporting Indian languages. It is an online multilingual, multimedia agricultural portal for disseminating information from and to the grassroots of the Indian agricultural community. The technology for Almost All Questions Answered (aAQUA) was developed by Developmental Informatics Lab, KReSIT, IIT B and was sponsored by Media Lab Asia and Development Gateway Foundation's R and D Center. aAQUA simultaneously addresses two major challenges in farmer outreach programs viz geographic reach and customized delivery. It answers farmers queries based on the location, season, crop and other information provided by farmers. An aAqua question is posted either by a registered user directly or through a telecenter/ kiosk operator who has an account in aAqua. Usually the question is from a farmer whose profile information provides details such as crop, farm size, pesticides and fertilizers used, dosage *etc.* The prices of various commodities along with their varieties are displayed spatially over a map. The user can decide where to sell his produce to get the maximum profit, depending on the prices and the distance of the markets.

Warana Wired Village

In the Warana Wired Village Project covering 70 villages in Maharashtra the existing cooperative structure has been used with state of the art infrastructure to allow Internet access to existing cooperative societies. The aim is to provide information to villagers by establishing networked booths in the villages. The villages in this sugarcane-growing region have computers that are linked to a central network that provides farmers access to essential pieces of information such as the ideal time for planting and harvesting sugarcane, the current market rates of their produce, and payments made by the factories. The Central and State governments together funded 90 per cent of the project.

The computer network has put an end to a major reason for anxiety at harvest time. Any delay in harvesting reduces its sugar content and, consequently, weight. Farmers are paid according to the crop's weight. The computer network provides each farmer with a share code. By punching the code into the system, the farmer gets details such as when the crop was planted and when it is due for harvesting. This gives the farmer sufficient time to organise workers to cut and transport the sugarcane.

Infosys' ICT Initiatives for Empowering Indian Farmers

Infosys Technologies has partnered with ACDI/VOCA, a non-profit international development organization that promotes broad-based economic growth, to develop an ICT-enabled application that would improve efficiencies in the agro supply chain in India. The solution successfully minimizes inventory requirements, reduces waste and allows retailers and farmers to be better integrated. This application falls under ACDI/VOCA's Growth-Oriented Microenterprise Development Program (GMED), which is a $6.3 million, USAID-funded initiative. GMED is an innovative program that develops sustainable and scalable approaches to job creation by fostering the growth of micro and small enterprises. Maintaining on-time, programmed delivery of fresh produce from a large and scattered production base is a complex and critical operation. This solution gives the organized retail sector access to a reliable small holder production base. It thereby decreases farm-to-market losses, currently estimated at 30 per cent to 40 per cent on certain products."

Metrological Information by Ingen Technologies

Many farmers in Punjab and West Bengal are receiving messages on their cell phones about weather information specific to towns and districts and by 2009 these could be availed by farmers throughout India. Offered by a Kanpur-based company Ingen Technologies, the service updates farmers on temperature, humidity and rainfall with additional parameters such as atmospheric pressure, solar radiation, wind speed and soil moisture. The system is approved and certified by the Indian Meteorological Department. The company is also offering its services to a major soft drink company, which can better predict demand for its beverages based on these predictions and analytics software. For farmers Ingen provides agro-advisory services that include advice on sowing times, disease outbreaks and frost forecast, through SMS. On other hand, Ingen has designed a decision support tool for utility companies and FMCGs, and have already supplied to some and are in talks with some other players.

Honey-Bee Knowledge Network

ICT can help empower the knowledge rich but economically poor people. Under the "**Honey-Bee" knowledge network** (of the IIM, Ahmedabad) implemented with the support from InfoDev division of World Bank the purpose is to augment grassroots inventors and overcome language, literacy and localism. The project has mobilized those creative and innovative farmers, artisans, mechanics, fishermen and women and labourers who have solved the problems through their own genius without any outside help, whether from state, market, or even NGOs. Such self triggered and developed innovations whether technological or institutional are scouted, supported, sustained and scaled up wherever possible with or without value addition, or linkage with formal science and technology institutions. Idea is to generate incentives and benefits for the innovators and traditional knowledge holders. The objective of this entire exercise was to create a clearing house, so that potential investors, venture capital or angle investors and entrepreneurs can link up with grassroots innovators, thus facilitating a golden triangle of innovation, investment and enterprise and thus build a bridge between formal and informal science.

ikisan Portal

The Nagarjuna Fertilisers Company Limited (NFCL) is an agribusiness based at Andhra Pradesh. They are disseminating various farming information to the farmers at various places through ICT centers. Ikisan is a comprehensive agri portal addressing the information, knowledge and business requirements of various players in the agri arena such as farmers, trade channel partners and agri input/output companies. ikisan provides online, detailed content on - crops, crop management techniques, fertilisers and pesticides and a host of other agriculture related material. ikisan enable farmers to network with other farmers, suppliers and consumers across the world.

Conclusion

Agricultural extension can play an important role in development. The public goods character of much extension work underpins the extensive public investment in extension services. But although public extension organizations are common in developing economies, they are often inadequately funded and their effectiveness is limited by many administrative and design deficiencies and challenges. Chief among these are the large scale and complexity of extension operations, the important influence of the broader policy environment, weak links between extension and knowledge generation institutions, difficulties tracing extension impact, problems of accountability, weak political commitment and support, the frequent encumbrance of extension agents with public duties beyond those related to knowledge transfer, and severe difficulties of fiscal unsustainability.

The evolutions and availability of ICT's has been the greatest communications revolution in recent years. Farmers now need information about trend and technology needed in farming so as to produce more and participate effectively in setting price of their product. To make all this possible huge utilization of ICT must be taken as the first priority. Although this is a development issue, it is just not the government, non-government organizations or the rural masses that have a role to play. Private profit-making institutions can develop solutions to capture the hitherto unrecognized markets, make profits and at the same time aid the rural societies. The new technologies being developed can help surmount barriers present in providing information resources at a low cost and make applications feasible and profitable.

Strategies for Effective Use of ICTs in Agricultural Extension

"The tools of ICT will provide networking of Agriculture Sector not only in the country but also globally and the Centre and State Government Departments will have reservoir of databases" and also "bring farmers, researchers, scientists and administrators together". The State's role in achieving ICT revolution becomes crucial if we examine the prospect of a sharply widening digital divide within the economy. Even beginning to provide access to the new technology to the overwhelming majority who cannot access it for technological reasons would impose a large financial burden. But the more difficult task is to prepare the disconnected to develop the competence to participate, however marginally, in the emerging digital economy. With literacy and schooling achievements still at indefensibly low levels, the first task of the government would be to rapidly advance the pathetic reach of literacy and school education in the country. In terms of priority this should be placed above the target of providing a minimum degree of access to ICT those who are completely disconnected. The application of ICT solutions for the development of rural India and other developing countries will surely open up a vast range of possibilities. Giving an opportunity to the vast majority of the population living in rural areas, to cross the digital divide to obtain access to information resources and services provided by ICT is the next revolution waiting to happen.

References

Ballantyne, P. and Bokre, D. (2003) *Report from the 'PrepCom'* (for CTA's Sixth Consultative Expert Meeting of its Observatory on ICTs). Wageningen, the Netherlands: CTA.

Batchelor, S. and O'Farrell, C. (2003) Guiding principles for ICT interventions. In: FAO (ed.) *Revisiting the 'Magic Box': Case Studies in Local Appropriation of Information and Communication Technologies (ICTs)*. Rome: Food and Agriculture Organization.

Dinar, A., and G. Keynan. 2001. "Economics of Paid Extension: Lessons from Experience in Nicaragua." *American Journal of Agricultural Economics* 83(3): 769–76.

Feder, G., A. Willett, and W. Zijp. 2001. "Agricultural Extension: Generic Challenges and the Ingredients for Solutions." In S. Wolf and D. Zilberman, eds., *Knowledge Generation and Technical Change: Institutional Innovation in Agriculture*. Boston, Mass.: Kluwer.

Feder, G., R. Murgai, and J. B. Quizon. 2004. "Sending Farmers Back to School: The Impact of Farmer Field Schools in Indonesia." *Review of Agricultural Economics* 26(1): 45–62.

Fleischer, G., H. Waibel, and G. Walter-Echols. 2002. "How Much Does It Cost to Introduce Participatory Extension Approaches in Public Extension Services? Some Experiences from Egypt." A case study prepared for the workshop on Extension and Rural Development: A Convergence of Views on Institutional Approaches, convened by the World Bank and the U.S. Agency for International Development, Washington, D.C. 12–15 November 2002.

Food and Agriculture Organization and World Bank. 2000. *Agricultural Knowledge and Information System for Rural Development: Strategic Vision and Guiding Principles*. Rome.

Greenridge, C. (2003) Welcome Address: ICTs Transforming Agricultural Extension? Presentation to CTA's Sixth Consultative Expert Meeting of its Observatory on ICTs. Wageningen, the Netherlands: CTA.

Van den Ban, A. W., and H. S. Hawkins. 1996. *Agricultural Extension*, 2nd ed. Oxford: Blackwell.

Feder, G., A. Willett, and W. Zijp. 2001. "Agricultural Extension: Generic Challenges and the Ingredients for Solutions." In S. Wolf and D. Zilberman, eds., *Knowledge Generation and Technical Change: Institutional Innovation in Agriculture*. Boston: Kluwer.

Feder, G., R. Murgai, and J. B. Quizon. 2004. "Sending Farmers Back to School: The Impact of Farmer Field Schools in Indonesia." *Review of Agricultural Economics* 26(1): 45–62.

Fleischer, G., H. Waibel, and G. Walter-Echols. 2002. "How Much Does It Cost to Introduce Participatory Extension Approaches in Public Extension Services? Some Experiences from Egypt." A case study prepared for the World Bank in Extension and Rural Development: A Convergence of Views on Institutional Approaches [illegible]

[illegible]

Chapter 17

Hydroponics (Soilless Culture): A Breakthrough Innovation in Agriculture

G. Dileep Kumar[1], M. Dhivya[2], M. Kokila[3]

[1,2]Research Associate, Department of Seed Science and Technology,
[3]Research Associate, Department of Floriculture,
Tamil Nadu Agricultural University,
Coimbatore – 641 003, Tamil Nadu

Introduction

Hydroponics is often defined as "the cultivation of plants in water." Hydroponics is however a technique for growing plants without using soil. Utilizing this technology, the roots absorb a balanced nutrient solution dissolved in water that meets all the plants developmental requirements. Research has determined that many different aggregates or media can support plant growth, therefore, the definition of hydroponics has been broadened to: **"the cultivation of plants without soil."**

Advantages of Hydroponics

1. The possibility of obtaining more products in less time than using traditional agriculture:

2. The possibility of growing plants more densely
3. Possibility of growing the same plant species repeatedly because there is no soil depletion
4. Plants have a balanced supply of air water and nutrients
5. More product/surface unit is obtained
6. Cleaner and fresher products can be reaped
7. Production can be timed more effectively to satisfy market demand
8. Healthier products can be produced
9. Products are more resistant to diseases
10. Natural or Biological control can be employed
11. Soil borne pests (fungi) and diseases can be eliminated
12. Troublesome weeds and stray seedlings which the result in the need for herbicides use and increase labour cost, can also be eliminated
13. Reduction of health risks associated with pest management and soil care
14. Reduced turnaround time between planting as no soil preparation is required
15. Stable and significantly increased yields and shorter crop maturation cycle
16. Can be utilized by families with small or no yard space
17. When water is used as the substrate:
    a. No soil is needed
    b. The water stays in the system and can be reused - thus, lower water costs
    c. It is possible to control the nutrition levels in their entirety - thus, lower nutrition costs
    d. No nutrition pollution is released into the environment because of the controlled system
18. Pests and disease are easier to get rid of because of container mobility.

Disadvantages of Hydroponics

1. Commercial Scale requires **technical knowledge** as well as a good grasp of the principles
2. On a commercial scale the initial investment is relatively high
3. Great care and attention to detail is required, particularly in the preparation of formulas and plant health control
4. A constant supply of water is required

Production System

Hydroponics can be classified as:

1. Open system
2. Closed system

Open System

In the open system of hydroponics, the nutrient solution is mixed and applied to the plant as required, instead of being re-cycled. Examples of some open system are:

1. Growing beds
2. Columns made out of tubular plastics or vertical and horizontal PVC pipes
3. Individual containers e.g. pots, plastic sacks and old tires

Closed System

In this system the nutrient solution is circulated continuously, providing the nutrients that the plant requires. Examples of closed systems include:

1. Floating roots
2. Nutrient Film Technique (NFT)
3. PVC or bamboo channels
4. Plastic or polystyrene pots set up in columns

Major Requirements that a Hydroponics System

Must Satisfy

- ☆ Provide roots with a fresh, balanced supply of water and nutrients
- ☆ Maintain a high level of gas exchange between nutrient solution and roots
- ☆ Protect against root dehydration and immediate crop failure in the event of a pump failure or power outage

Location

Constructing a Container

If there is a need to build a container, you may consider building a bed or box with the following size:

- ☆ Length : 1.25m
- ☆ Width : 0.95m
- ☆ Depth : 0.10 m

The materials needed to build the box or bed are:

- ☆ Wood : 15 feet of 1 x 4 inch wood shingle or plank
- ☆ 166 feet of ½ x 3 inch wood shingle or plank
- ☆ Black plastic : 5.6 feet x 4.3 feet
- ☆ Nails : 1 pound – 2 inch

The tools and materials needed to build a bed are: hammer, saw, meter rule, stapler, staples, drill, drill bits, level, saran netting, water hose and water.

The characteristics of a good substrate:

- ✰ It must be made of particles no larger than 7mm and no smaller than 2mm
- ✰ It must be capable of maintaining moisture and draining excess liquid
- ✰ It must not degrade or decompose easily
- ✰ It must not hold microorganisms hazardous to human or plant health
- ✰ It must not be contaminated with industrial residual waste
- ✰ It must be readily available
- ✰ It must be potable

Recommended Substrate Mixtures

Some recommended substrate mixtures are:

- ✰ 50 per cent rice hull : 50 per cent ground volcanic stones
- ✰ 60 per cent rice hull : 40 per cent sand
- ✰ 60 per cent rice hull : 40 per cent ground clay bricks
- ✰ 80 per cent rice hull : 20 per cent saw dust

Another substrate which could be used is:

- ✰ Clean rain water

Rice Hulls

These must be washed and kept very moist for ten (10) days in order that all seeds in the rice hulls will germinate. The germinated seedlings must be removed.

Saw Dust

Saw dust may be used in small quantities, 15-20 per cent of the substrate since large quantities are harmful to some plants.

Nutrients and Fertilizers

The hydroponic solution contains a balanced amount of nutrients to produce healthy and productive plants. In addition to the elements (carbon, hydrogen, oxygen) that vegetables extract from the air and water, plants need some elements that may be classified by quantities and need.

Future of Hydroponics

Hydroponics is a relatively new technology, evolving rapidly since its inception 70 years ago. From its origins in academic research, to its utilization in industry and government, hydroponics has found many new applications. It is a versatile technology, appropriate for both developing countries and high-tech space stations. Hydroponic technology can efficiently enerate food crops from barren desert sand

and desalinated ocean water, in mountainous regions too steep to farm, on city rooftops and concrete schoolyards and in arctic communities. In highly populated tourist areas where skyrocketing land prices have driven out traditional agriculture, hydroponics can provide locally grown high-value specialty crops such as fresh salad greens, herbs and cut flowers.

Like manufacturing, agriculture tends to move toward higher-technology, more capital-intensive solutions to problems. Hydroponics is highly productive and suitable for automation. However, the future growth of controlled environment agriculture and hydroponics depends greatly on the development of systems of production that are cost-competitive with those of open field agriculture. Improvements in associated technologies such as artificial lighting and agricultural plastics, and new cultivars with better pest and disease resistance will increase crop yields and reduce unit costs of production. Cogeneration projects, where hydroponic greenhouses utilize waste heat from industry and power plants, are already a reality and could expand in the next few years. Geothermal heat could support large expanses of greenhouses in appropriate locations.

It has been proposed that glasshouses located in deserts of the world could one day serve a dual purpose, where antenna could be embedded into the glass to receive energy radiation from an array of energy collectors in space, while at the same time facilitate hydroponic tomato production.

The economic prospects for controlled environmental agriculture and hydroponics may improve if governmental bodies determined that there are politically desirable effects of hydroponics that merit subsidy for the public good. Such beneficial effects may include the conservation of water in regions of scarcity or food production in hostile environments; governmental support for these reasons has occurred in the Middle East. Another desirable societal effect could be the provision of income-producing employment for chronically disadvantaged segments of the population entrapped in economically depressed regions; such employment produces tax revenues as well as personal incomes, reducing the impact on welfare rolls and improving the quality of life.

Hydroponics is a technical reality. Such production systems are producing horticultural crops where field-grown fresh vegetables and ornamentals are unavailable for much of the year. The development and use of controlled environment agriculture and hydroponics have enhanced the economic wellbeing of many communities throughout the world.

Commercial

Some commercial installations use no pesticides or herbicides, preferring integrated pest management techniques. There is often a price premium willingly paid by consumers for produce that is labelled "organic". Some states in the USA require soil as an essential to obtain organic certification. There are also overlapping and somewhat contradictory rules established by the US Federal Government, so some food grown with hydroponics can be certified organic. Most hydroponically grown produce cannot be sold as organic due to the fact that they do not use soil as a growing medium.

Hydroponics also saves water; it uses as little as 1/20 the amount as a regular farm to produce the same amount of food. The water table can be impacted by the water use and run-off of chemicals from farms, but hydroponics may minimize impact as well as having the advantage that water use and water returns are easier to measure. This can save the farmer money by allowing reduced water use and the ability to measure consequences to the land around a farm.

To increase plant growth, lighting systems such as metal-halide lamp for growing stage only or high-pressure sodium for growing/flowering/blooming stage are used to lengthen the day or to supplement natural sunshine if it is scarce. Metal halide emits more light in the blue spectrum, making it ideal for plant growth but is harmful to unprotected skin and can cause skin cancer. High-pressure sodium emits more light in the red spectrum, meaning that it is best suited for supplementing natural sunshine and can be used throughout the growing cycle. However, these lighting systems require large amounts of electricity to operate, making efficiency and safety very critical.

The environment in a hydroponics greenhouse is tightly controlled for maximum efficiency, and this new mindset is called soil-less/controlled-environment agriculture (CEA). With this growers can make ultra-premium foods anywhere in the world, regardless of temperature and growing seasons. Growers monitor the temperature, humidity, and pH level constantly.

Hydroponics have been used to enhance vegetables to provide more nutritional value. A hydroponic farmer in Virginia has developed a calcium and potassium enriched head of lettuce, scheduled to be widely available in April 2007. Grocers in test markets have said that the lettuce sells "very well", and the farmers claim that their hydroponic lettuce uses 90 per cent less water than traditional soil farming.

Advancements

With pest problems reduced, and nutrients constantly fed to the roots, productivity in hydroponics is high, although plant growth can be limited by the low levels of carbon dioxide in the atmosphere, or limited light exposure. To increase yield further, some sealed greenhouses inject carbon dioxide into their environment to help growth (CO_2 enrichment), add lights to lengthen the day, or control vegetative growth, *etc.*

Chapter 18

Enhancing Agricultural Productivity through Effective Dryland and Waste Management Techniques

B. Sivakami[1] and S.P. Ramanathan[2]

[1]Senior Research Fellow, Directorate of Extension Education
[2]Professor, Water Technology Centre
Tamil Nadu Agricultural University,
Coimbatore – 641 003, Tamil Nadu

Introduction

At present India experiences a deficit of 50 per cent green fodder and 25 per cent dry fodder. There exists a mismatch between supply, demand and fodder, which deserves immediate and prime attention. On the other side, vast stretches of land remain without green cover and are categorized either as degraded or unproductive. This underscores the need for developing a technology that can solve both these problems. The best solution would be developing permanent pastures and cattle rearing which is the promising and sustainable enterprise on degraded and poor sites as the demand for milk and meat is ever increasing.

Converting Barren and Wasteland into Productive Assets

The average annual rainfall in Tamil Nadu is about 911.60 mm and the average rainfall in northern and eastern districts, is above the average but in western and southern districts it is less than the state average. From the rainfall data analysed, it is seen that about 40-50 per cent of the state is getting more than 350 -400 mm during both the South West and North East monsoon seasons. The indicates that with proper water harvesting, land development and selection of suitable crops, it is possible to take 2 crops in at least about 30-35 per cent of these lands or take a tree crop/horticulture (fruit) tree. For this, the terrains have to be studied in details. Rainfall should be analysed critically and suitable crops/trees are t be identified in consultation with farmers/stake holders. Since both the Government of India/ Government of Tamil Nadu are expanding a log of money to develop these type of lands under DPAP, NAWPRA, WGDP and wasteland development programme, it is possible that the barren and uncultivated lands can be brought under plough which not only helps the poor farmers to get some yields from these wasted lands but also generate employment opportunities in the rural areas.

Increasing Agricultural Production in Drylands and Wastelands

Food production has to be increased from the present 240 M.T to 500 M.T. by 2050 for the expected population of 165 crorers in India. Since malnutrition is already prevalent in the country and food, food and feed production would have to increase more than double, during the period to adequately feed the population. In the last century most gains in food production have come from irrigated agriculture and from areas with favourable growing conditions, such as assured water (rain) and fertile soils.

There are indications that doubling of the food production may not come from the present high input production systems. The land area under irrigation has not increased in the last 3 decades in Tamil Nadu. Further, the rate of irrigated land going out of production because of soil degradation, salinity and alkalinity problems is rapid. These trends illustrate that the growth in food production in high potential environments and high input production system may level off or even decline in the coming years. Therefore in order to meet the increasing demands for food, pressure on rainfed agricultural lands is likely to increase and cultivation be extended to fallow and wastelands. If pressure on these lands increased further, degradation of the resource base may bad to the point where agriculture production becomes uneconomical and these lands may be abandoned by the rural population. The dry lands support a large number of rural populations in the country. If no additional measures are taken up to stabilize or improve the agricultural resource base in these areas, the sustainability of rural livelihoods will be at risk. Major and concerted effort is needed to reverse degradation of the resource base in these areas so that a sustainable production system is possible in these lands.

Agro Ecological Characterization of Dryland

Drylands are defined as those areas with fewer than 120 growing days/period whereas the arid and semi arid zone is defined as the area with a lengthof growing period of less that 180 days. The definition of dry and semi arid zone are both based on the concept of Length of Growing Period (LGP). The LGP during the year when rainfed available soil moisture supply is greater half the potential evapotranspiration (PET). It includes the period required to evaportranspire up to 100 mm of available soil moisture stored in the soil profile, rain water. However, some soil stores less or some store more than 100 mm. If a soil stores more than 100 mm (deep vertisoils) the effective LGP will be longer than the LGP estimated on the basis of a moisture storage capacity of 100 mm. If the soil stores less than 100 mm (shallow or sandy alfisols) the effective LGP will be less than the LGP estimated on the basis of a moisture storage capacity of 100 mm. similarly, if the water holding capacity exceeds the amount of water actually stored in the soil. And if the storage of moisture is lost due to surface runoff, surface evaporation or deep drainage. The upper limit of the LGP is 120 days in the case of dry lands and 180 days in the case of semi dry zones. In order to grow rainfed crops, there has to be al lower limit to the LGP as well. The lower limit is basically set by the minimum crops growth duration, *i.e.* the time taken for a crops to reach physiological maturity. For example, several short season varieties of cereal crops, such as millet and sorghum and pulses can reach maturity in 75 -90 days in the semi arid tropics. The LGP that separates arid from semi arid is thus set as 75 days by FAO. The semi arid zone refers to a LGP between 75 and 120 days. In regions with LGP of less than 75 days conventional crop production is not possible or is a very marginal activity with regular crop failures. Most of these dry or arid and are used as range lands with trees. It may be concluded that although the concept of LGP is very useful in dry and semi arid environments, one has to be cautious in interpreting LGPs for specific location. The design of cropping system and probabilities of amount and distribution of rainfall, long term rainfall trends, temperature regimes and soil conditions.

Crop Improvement and Production System

An indicated in the pervious chapters, there is a large gap between the yields at experimental stations and those in farmer's fields. In fact we might wonder why one should try in to increase the yield potential of improved varieties by another 5 to 10 per cent when the yield gap is of the order of 50 per cent, 80 per cent or more. Hence it may be more useful to concentrate on the question why there is such a large gap between yield potential and actual yield in farmer's field.

Basically, the only technology that has been successfully transferred to the farmers' field is improved seed plus a relatively simple high input technological packages, like early planting and the use of fertilizer and biocides. Farmers will be eager to adopt these technologies if the necessary inputs are made available and supported by adequate facilities by the government. The main reason for the yield gap is that improved varieties were not selected for those environments and they do not perform well in farmer's field as they performed in experimental stations. Further, the agro climatic conditions at the experimental stations may be different well without a resource management plots. The soil and water conservation practices may be very important in the farmer's field to realize the potential crop yield. Similarly all other practices like weed control, control of pest and diseases, application of nutrients, date of planting *etc.* should begiven importance. Another important reason is inadequate extension services and inputs such as improved seed, fertilizers and credit facilities to the farmers.

High and Low Potential Production Environments

To assess the opportunities for transferring improved technologies to farmers for increased food production, it may be useful to distinguish explicitly high potential production environments and low potential production environment differs from low potential production environments in soil, climate and socio economic conditions. Soils in high potential production environments are generally better in deep soils in that they are flat they are flat and heavy textured with high water holding capacity. Soils in low potential production environments are marginal, light to medium textured in sloppy land and in shallow depth. Climatic conditions are more favourable high potential production environments than in low potential production environments. Therefore, crop improvement and resource management research have to take into consideration the difference between high and low potential production environments and high and low input production systems.

Technologies for Low Potential Production Environments

The International Agricultural Research Centre (IARC) has made more impact in high potential production than in low potential production environments. The reason is that the IARC have not been very successful in transferring resource management technologies suitable for low potential production environments. Further the impact was higher, adoption was easier and technologies were simple for high potential areas. The technologies suitable for low potential production environments have not been developed as the scientists concentrate their research at the experiment stations rather than in farmers fields and further they often do not involve the farmers in design and management of the experiments other than land preparation.

Intercropping of Grasses + Legumes under Rainfed Conditions

Field investigation has been done at Pudukkottai to develop grass –legume mixtures for establishing permanent pastures in dry areas, that have capacity to maximize fodder production in a sustained manner besides keeping fertilizer nitrogen requirement to a minimum and crude protein yield to a maximum.

Among the grass fodders *Andropogangayanus* is an extremely drought tolerant perennial grass capable of surviving rainless spells fro even five months. Cenchrus species (Cenchurssetigerus, Cenchrusciliaris) calles as "Kozhukattai pull" in Tamil, characterized by tolerant to severe drought, high temperature and suited for marginal lands. Clitoria ternate is one of the best – suited legume species being grown under rain fed conditions. Stylosanthuscabra (Muyalmasal) was introduced in India during 1970's as a perennial herb with 19 per cent cured protein content. Desmanthus virgate (Velimasal) commonlyknown as Hedge Lucerne obtained from Thailand. It can grow up to a height of about two meters amenable for multiple cutting.

Field experiments were conducted in alfisol with a pH of 6.0 during 2000-2001 and 2001-2002 at National Pulses ResearchCentre,Vamban, Pudukkottai. The fertility status of the soil was low in available N (1925kg/ha) and medium in available P (14.1kg/ha) and available K (151.0kg/ha). This centre receives an annual rainfall of

930mm (average of 25 years? 34 rainy days. The mean temperature ranges from 24.5 c (minimum) to 35 C (maximum). Three fodders *viz., Cenchrussetigerus, Cenchursciliaris, Andropogangayanus* and three planting ratios (3:1, 4:1 and 5:2) with three intercrops Desmanthus virgate, Clitoria ternate and Stylosanthusscabra have been tested. These intercrops *viz.,* Desmanthus virgate, Clitoria ternate and Stylosanthusscavra gave in crude protein content of 19.2, 21.3 and 19.0 per cent respectively, from the above study it can be inferred that under drought period Andropogangayanus (Crude protein – 19.5 per cent) and Desmanthus virgate (Crude protein – 19.2 per cent) tolerated rainfed conditions of Pudukkottai district and yielded 13.5t/ha/yr of nutritious green fodder.

Intercropping of Ragi + Pulses under Rainfed Conditions

Ragi is traditionally grown as Salkeppai under rainfed condition in Tamil Nadu. Being a close spaced crop intercropping seems t be an uphill task to the farmers. Field investigations has been done with an objective to develop a ragi based intercropping system with short duration pulses under rainfed situation of Pudukkottai district.

The result reveal that maximum enhancement in growth and yield attributes have been realized when ragi (CO13) was intercropped with green gram (Vamban 2) compared to black gram (Vamban 3) or red gram (Vamban 1) with regard to row ratio, adoption 2:2 row ratios proved significantly superior to 4:2 and 6:2 ratios. Equal spatial arrangement of ragi and pulses resulted in better utilization of space, moisture and applied nutrients. The 2:2 ratios seems to have better complementary effect betweenragi and green gram (Vamban 2). The ragi grain equivalent yield and land equivalent ratio markedly superior under intercropped situation that that of sole cropped one. Among the intercropping systems tested, ragi + green gram recorded higher ragi grain equvatent yield (2691kg/ha), higher land equivalent ratio (1.56) and higher benefit cost ratio (3.03) followed by ragi + black gram.

It is evident from the study that ragi + green gram (Vamban) at 2:2 ratio can economically and profitably be accommodated under rainfed condition for better yield advantage and enrichment of soil health.

920 mm (average of 25 years) 34 rainy days. The mean temperature ranges from 24 °C (minimum) to 35 °C (maximum). Three fodder crops, Cenchrus setigerus, [illegible], Andropogon gayanus and three planting rates (50, 40 and 30) with three intercrops Desmanthus virgatus, Clitoria ternatea and Stylosanthes scabra have been tested. These intercrops viz., Desmanthus virgatus, Clitoria ternatea and Stylosanthes scabra gave crude protein content of 19.2, 21.3 and 18.1 per cent respectively. From the above study it can be inferred that under drought period, Andropogon gayanus (Crude protein = 14.5 per cent) and Desmanthus virgatus (Crude protein = 19.2 per cent) tolerated extended conditions of [illegible] and yielded [illegible] of nutritious green fodder.

Intercropping of Ragi + Pulse under Rainfed Conditions

Chapter 19

Innovative Agriculture for Sustainable Livelihood of the Farmers

A. Anitha Pauline[1] and C. Karthikeyan[2]

[1]Assistant Professor, Sethu Bhaskara Agricultural College and Research Foundation,
[2]Professor (Agrl. Extension), e-Extension Centre Tamil Nadu Agricultural University, Coimbatore – 641 003, Tamil Nadu

---

Introduction

The purpose of this chapter provokes discussion by exploring and elaborating the concept of sustainable livelihoods. It is bases normatively on the ideas of capability, equity and sustainability each of which is both end means.

In the 21st century livelihoods will be needed by perhaps two or three times the present population. A livelihood comprises people their capabilities and their means of living, including food, income and assets. Tangible assets are resources and stores and intangible assets are claims and access. A livelihood is environmentally sustainable when it maintains or enhances the local and global assets on which livelihoods depend and has net beneficial effects on other livelihoods. A livelihood is socially sustainable which can cope with and recover from stress and shocks and provide for future generation.

With about 35 to 40 per cent of the population living in poverty, livelihood security for the rural poor continues to be a cause of concern in India. Indian economy is heavily dependent on agriculture even today because about 65 per cent of the population is living in rural areas and over 80 per cent of them are dependent on agriculture and allied activities for their livelihood. Out of the total 129.22 million land holders in the country, 64.8 per cent are marginal holders who own less than one hectare and 18.5 per cent families are small farmers owning between one and two hectares. More than 50 per cent of these families are located in arid and semi-arid regions, where the rainfall is scanty and erratic. These farmers have been growing drought tolerant food crops, mostly millets and pulses with very low investment in improved seeds, fertilisers and plant protection measures, resulting in poor yields and low returns. Fragmented land holdings, heavy depletion of soil productivity, inefficient use of water resources, out-dated agricultural production technologies, unavailability of agricultural credit and lack of infrastructure for post harvest management and marketing of agricultural produce, are the other factors which further suppress their agricultural production. Unfortunately, these regions have also been neglected by the scientific and business communities in introducing new technologies, high yielding varieties which are resistant to drought and developing necessary infrastructure as well as support services to boost agricultural production and value addition. Due to low agricultural productivity, these small and marginal farmers as well as about 15 to 18 per cent landless families living in rural areas are unable to generate remunerative employment and about 40 per cent families are forced to live in poverty.

For these small holders and landless, livestock has been a source of supplementary income. However, over 75 per cent of the animals are uneconomical due to severe genetic erosion, inadequate feeding and poor veterinary care. With lower crop and livestock productivity, the employment opportunities in the farming and other related sectors are reduced further, leading to reduction in farm wages, seasonal employment, malnutrition and migration. With lack of food security, poor families are compelled to migrate to cities in distress, keeping their agricultural lands fallow. Such barren lands accelerate soil erosion, run off of rain water, resulting in floods, siltation of water bodies and loss of biodiversity and thereby contributing to global warming. In the absence of efficient soil and water conservation, there will be a severe reduction in the ground water table, accelerating the process of denudation of the eco-system and shortage of drinking water. Distress migration will also deprive the women and children of their basic needs such as shelter, safe drinking water and health care, which will affect their quality of life. The children will discontinue their education and end up as child labour and illiterate unemployed youth of the future. Thus, improving the agricultural productivity of small land holders can play a key role in ensuring food security and improving the quality of life in the country.

Scope for Technology Transfer

Rural population in India has been facing series of problems which affect their progress and quality of life. Most significant among these problems are lack of gainful employment leading to food insecurity, illiteracy and poor health. Although agriculture contributes only 30 per cent of the Gross National Product (GNP) of the country, the rural economy is heavily dependent on agro-based activities. This is primarily due to lack of industrial infrastructure and limited opportunities for providing jobs in the service sector. It has been observed that over 85 per cent of the rural people are dependent on agriculture and most of them are engaged in the struggle for food security. Therefore, it is necessary to promote suitable rural technologies which can enable the local communities to enhance their efficiency and earnings. With the improvement in agricultural production, various opportunities can emerge in the agri-business and non-farm sectors in the future.

There are many Governments and Non-Government Agencies engaged in providing sustainable livelihood to the rural poor. Over the years, a large number of activities have been identified both in On-Farm and Non-Farm sectors.

Technologies for Rural Livelihood

A. On-farm Activities

- ✰ Crop production
- ✰ Horticulture, Forestry, Sericulture
- ✰ Livestock husbandry, fishery
- ✰ Agro Service Centres
- ✰ Processing of Food and Forest Products
- ✰ Production of Agricultural Inputs: Biofertilisers, Biopesticides, Vermicompost, Mushroom spawn production, Seeds and Plants, Cattle feed

B. Off-farm Activities

Cottage Industries: Pottery, Smithy, Carpentry, Textile, Production of building materials.

Services: Automobile hire and repairs, Electrical works, civil construction, Consumer stores.

While promoting these technologies, it was observed that the success of technology transfer is dependent on various factors, particularly the infrastructure for providing motivation, training, finance, processing and marketing. As our target groups, who are semi-literate and economically backward, it is difficult for them to search for appropriate technologies on their own for enhancing their income. There is a need for facilitating organizations which can identify various technologies and modify them to suit the local needs before transferring them to the beneficiaries.

Various other technologies for income generation and appropriate technologies are also essential to promote hygiene, sanitation, supply of clean drinking water and community health. Information and communication technologies are needed to reduce the gap between the urban and rural communities and empower them.

Replicable Technologies for Sustainable Livelihood

The relevant sustainable technologies are including poultry and fishery primarily because of its benefit to a large number of small and poor farmers. The other relevant technologies for the rural poor are water resource management, wastelands development, sericulture, agri- horti- forestry, eco-friendly agriculture, mushroom cultivation, vermicomposting, *etc.* All these technologies can help the farmers to earn adequate income to come out of poverty. In poor situation, it is necessary to promote multidisciplinary activities which can help the farmers to general substantial income from different sources. Important technologies for providing sustainable livelihood in rural areas are presented below:

1. *Livestock Development*

Dairy husbandry was considered as the most powerful tool, as most of the cattle and buffaloes in the country inspite of being low productive, could still be used as breeding stock to produce superior quality progeny. Timely advice on feeding, fodder production, health care, deworming, use of mineral mixture, *etc.* has enabled farmers to increase milk production. A family with 3 crossbred cows, each producing 2200–3000 litres of milk/lactation, can generate a net income of ₹30,000– ₹40,000 annually.

Goat rearing with strict control on open grazing, also has good potential. There are a large number of small farmers, landless and women-headed families, who are dependent on small species of livestock such as goat, sheep, pigs, poultry, *etc.* for their livelihood. Goat husbandry is more popular because of low risk, short gestation and high returns. A family maintaining 6-8 goats can earn an annual income of ₹ 12,000–15,000 per annum. In the regions where livestock husbandry is remunerative, there are new business opportunities in the form of small scale industries for production of concentrate feeds, mineral mixture and complete feed which can facilitate even the landless to depend on livestock husbandry. Some of these units need significantly high capital investment in the initial stages. Hence, local NGOs or cooperatives can be encouraged to establish such units with support from farmers and financial institutions. There is an excellent opportunity for local farmers' groups to establish small scale dairies in small towns and block headquarters, which can handle 5000–20000 litres of milk per day to produce a wide range of products apart from supplying pasteurised milk. Such dairies can be efficient and reduce the number of middlemen in the market, while enhancing the demand for milk products.

2. *Wastelands Development*

Large areas mostly owned by the communities, are presently underutilized due to low productivity. Such lands which were sources of fodder, fuel and timber in the past, have now turned into wastelands, posing a serious threat to the water table, eco-system and environment. The wastelands provide an excellent opportunity for generating income for the local communities, particularly the landless and the small farmers. The development of community lands depends on the soil productivity, water availability and the local needs. While fertile lands with assured source of water can be brought under multipurpose tree species of fruits, nuts, medicinal herbs, *etc.*, less productive barren lands can be brought under silvipasture. With an investment of ₹12,000–15,000 per ha for the development of community pasture, out of which 85-90 per cent was on labour wages, the pasture started producing fodder worth ₹ 5000–7000/ha/year, which was shared by the members of the community.

The programme could help the poor and landless to take up livestock husbandry as the main source of livelihood.

3. *Water Resource Development*

As only 35–40 per cent rain water is effectively used and over 70 per cent of the agriculture is dependent on rainfall, there is good scope to promote water resources development particularly in arid and semi-arid regions in the country. Efficient rainwater harvesting along with promotion of sustainable agricultural practices, can improve the cropping intensity by 25-30 per cent and increase the crop yield by 40–70 per cent while generating employment throughout the year.

4. *Tree-based Farming*

In rainfed areas, where there are many constraints for increasing the crop yields, tree-based farming can be a boon as trees have better ability to tolerate erratic weather conditions. Depending on the soil fertility and water availability, different tree species can be grown. While horticulture is most payable, timber and round wood plantations are also profitable compared to rainfed farming. Horticulture is ideal for small holders as they can effectively utilize their idle labour while round wood and pulp wood plantations will be good for large holders where most of the operations can be mechanized.

5. *Vermicomposting*

Small farmers and landless can take up vermicomposting as an income generation activity. They can collect organic matter either from community lands and forests, procure inferior quality crop residues and weeds, and collect dung to initiate vermicompost production on a commercial scale. This is a popular activity preferred by women in many states. There is good demand for the produce even by local agriculturists because of its beneficial effects.

6. *Production of Biofertilisers and Biopesticides*

Farmers either individually or in groups can take up the production of biofertilizers, biopesticides and biofungicides, using locally available inputs. This will not only generate rural employment but also enable farmers to take up organic farming and reduce the cost of production.

7. *Sericulture*

As mulberry plants can be established as intercrop in fruit orchards, sericulture has good potential to provide substantial income within a short gestation of 4-6 months. Mulberry can also be grown as a sole crop on marginal lands. Farmers maintaining mulberry on one ha of land for rearing silkworm, can generate a net income of ₹0.8 to 1.2 lakhs per year.

8. *Mushroom Cultivation*

Cultivation of Oyster mushroom (*Dhingri*) has been promoted a large scale. A family with a shed of 3 x 3 m to grow mushroom, can earn a net income of ₹5000–₹8000 annually through 6-8 cycles. However, assured market is a critical factor for the success.

9. *Food Processing*

It is estimated that about 20-30 per cent of the agricultural products, particularly fruits and vegetables, are lost due to poor handling. Exploitation by the traders further affects the profit. Hence, processing of horticultural produce is not only essential but also provides an opportunity to earn additional income.

C. Non-Farm Activities

There is good scope for promotion of various income generating activities in the rural non-farm sector. These include the following:

a. Training of youth in masonry, carpentry, smithy, repairs of cycles and motor cycles, tractors, pumpsets, electrification and winding of motors, *etc.* Subsequent to training, entrepreneurs need working capital and critical equipment to start their services.
b. Rope making and mat making by using locally available agricultural by-products and grass
c. Bamboo crafts and utility articles
d. Production of housing materials
e. Embroidery and tailoring
f. Flour mills and oil seed expellers
g. Establishment of grocery shops and food stalls.

Strategies for Sustainable Development

I. Family as a Unit for Development

It is necessary to consider each poor rural family as the basic unit of development. This provides an opportunity to identify the target families who require different types of support to come out of poverty. Generally most of the community development programmes consider village as the unit of development where the well to do and influential sections of the society are likely to dominate over the poor and exploit the benefits to the maximum extent. Thus, such development projects may often create a wider gap between the rich and poor within the community.

II. Focus on Quality of Life

The overall goal should be to ensure better quality of life through promotion of various development activities related to livelihood, health, literacy and moral development. Food security being the most serious concern of poverty, livelihood programme should be given priority. Hence, it is necessary to blend livelihood programmes with education, health care and moral development activities. These components are generally acceptable to all the members of the community, irrespective of their religious and ethnic backgrounds which can bring significant change in the attitude of the target communities.

Assured Livelihood

Development programmes should have a primary goal of helping the poor families to come out of poverty within the shortest period. The dairy development programme has a gestation period of 3-4 years till the newly born calf comes into milk production. In land-based development programmes, the gestation period may vary from 2 to 6 years, depending on the type of farming systems practiced by the farmers. In case of arable crop production, the gestation period is small due to short rotation crops while the fruit and tree crops take 5-6 years to generate income.

The other important aspect is to provide support during the gestation period. Many of the poor who do not have any resources even to procure their daily ration, are likely to neglect their development work, if no support is available in the form of assistance or wages to ensure their food security. Hence, different short term income generation activities need to be incorporated to ensure that farmers are able to generate some income from other sources till the income starts from the major interventions.

III. Women Empowerment

Involvement of women in all the development programmes right from the stage of project planning is essential. Although women represent 50 per cent of the population, their contribution to the development of the nation is even more significant as they also have the important responsibility of grooming children and procuring the basic needs required for food, fuel and fodder securities. Active participation of women in development programmes will help to identify their problems and reduce their drudgery. It is difficult to make progress in agriculture without empowering women.

IV. Environmental Protection

In all the development programmes, conservation of the natural resources and protection of the environment are essentially built in, as these are critical for sustainable development. This is particularly important while dealing with the poor as their primary objective is to earn their livelihood and the development organisations have the obligation of carefully designing the programme to ensure environmental protection along with income generation activities.

V. Blending Development with Research and Training

For effective implementation of various development programmes, the development programmes are supported by applied research and training activities. It has been realised that any development programme without research back up is outdated and any research programme without development and extension outlets is academic. Training of the field functionaries and farmers is essential for effective transferring of technologies from laboratories to the field.

VI. People's Organisations

To sustain the benefits of various projects particularly after the completion of the project, strong infrastructure at the grassroot level is essential. Therefore, development of grassroot level People's Organisations, right at the initiation of the project will be helpful. Several types of local People's Organisations such Self Help Groups (SHGs), Village Level Planning Committees, Users' Groups of various goods and services, Networks and Federations of SHGs, Village Level Organisations and processing and marketing Cooperatives are some of the organisations which can be promoted in the field. Subsequently, they can work closely with the Panchayat Raj Institutions to participate in various state-sponsored development activities as well as to ensure the welfare of their community.

Sustainability

Proper planning is essential right from the inception of the programme to build strong grass root level people's organisations to manage the programme with least dependence on outsiders. The next step is to shift from the role of a development organisation to a service provider, where the participants demand various services on payment of reasonable fees, instead of accepting whatever is given to them. This helps in sustainability of the programme, beyond the project period.

Conclusion

The future with policy support and adequate funding, sustainable agricultural could be intensified over large production areas in a relatively short period of time. The challenge facing policymakers is to find effective ways to scale up sustainable approaches so that hundreds of millions of people today and tomorrow can benefit. Some of the key needs are:

- ✰ Large-scale investment in agricultural research to find out what works, where it works and how to adapt it to local contexts.
- ✰ Assessment of the harm caused by current practices on agro ecological systems.
- ✰ Decisions at national level about which production systems are unsustainable and which sustainable approaches are suitable for scaling up.
- ✰ Work with farmers to validate and adapt approaches to local ecosystems.
- ✰ Preparation of plans for investment in appropriate policies and institutions, including farmers' organizations.
- ✰ Monitoring, evaluation and review of progress, making adjustments where appropriate.

Throughout the world, it is clear that sustainable agriculture is the answer. Farmers have produced countless examples of the advances that result when people work in harmony with nature. In the long term, this is the only way we can achieve sustainable solutions to hunger and poverty.

References

1. Anon. 1996. National Livestock Policy Perspectives: Report of Steering Group. Planning Commission, Government of India.
2. Anon. 2005. Annual Report 2004-2005. Department of Animal Husbandry and Dairying, Ministry of Agriculture, Government of India. New Delhi.
3. Government of India, 2011. Faster, Sustainable and More Inclusive Growth – An Approach to the XII Five Year Plan, Planning Commission, New Delhi.

4. Anon. 1997. Dairy India Yearbook VI. New Delhi.

5. Blummel, M., Krishna, N. and Orskov, E.R. 2001. Supplementation strategies for optimizing ruminal carbon and nitrogen utlisation: Concepts and approaches. Proceedings of the 10th Animal Nutrition Conference: Karnal, India. Review Papers, November 9-11: 10-23.

6. Hegde, N.G. 2010. Mitigating Global Warming while providing Sustainable Livelihood through Integrated Farming Systems: Experiences of BAIF. International Conference on Global Warming: Agriculture, Sustainable Development and Public Leadership. Ahmedabad, India. Mar 11-13: 16 pp.

Chapter 20

Seed Quality Analysis through Advance Instruments

S. Sathish[1] and V. Vijayalakshmi[2]

[1]Research Associate, [2]Assistant Professor
Department of Seed Science and Technology,
AC and RI, Kudumiyanmalai, TNAU, Tamil Nadu

The present-day seed market is very competitive; farmers require seed lots with a germination capacity close to 100 per cent and high vigour. At the same time, one of the problems for seed producers is the lack of reliable and simple technology that can predict after-sowing seed performance and which are faster than assessing germination and vigour using laboratory tests. Such technology could also be used for monitoring seed priming, a pre-sowing treatment commonly used for the enhancement and improvement of germination. In the present scenario, advanced instruments have been developed to analyse the quality of seeds and to monitor the seed enhancement protocols in order to obtain desirable enhancement based on their vigour status. Following are the latest instruments developed for simplifying the seed quality analysis.

Flow Cytometry

Flow cytometry is a fast and accurate method for estimating nuclear DNA content. It involves preparation of aqueous suspensions of intact nuclei, the DNA of which is stained using a DNA-specific fluorochrome. The result of the analysis is usually displayed in the form of a histogram of the relative fluorescence intensity, representing relative DNA content (Figure 20.1). The method can be used to study the cell cycle in seeds by isolating nuclei from the tissues of the embryo and/or endosperm. The information obtained can be used for determining the physiological state of a seed during its development, maturation, processing and germination, and therefore can be useful for monitoring the production of high quality commercial seed lots.

Instrumentation

The key components of a typical analytical flow cytometer include fluidic system, excitation light source (lasers), optics, electronic detectors, amplifiers, analog-to-digital converters, and pulse processors (Figure 20.1).

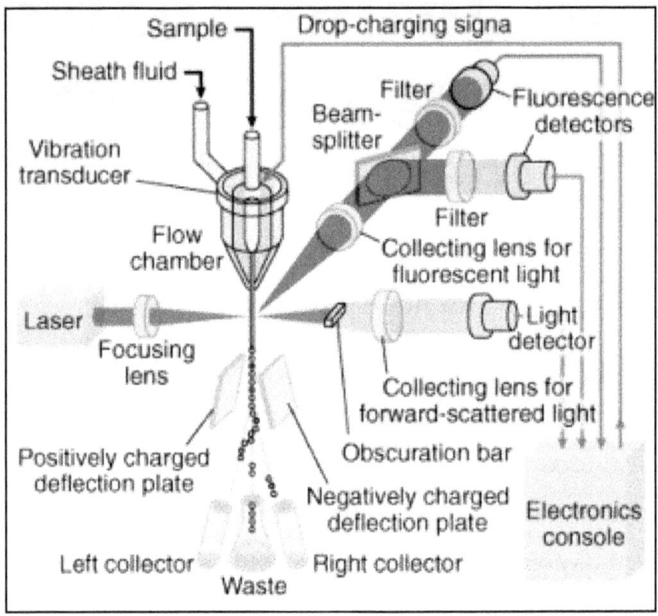

Figure 20.1: Components and Working Principle of Flow Cytometer.

Flow Cytometer Operation

The prepared sample of particles under investigation begins its journey through the ?ow cytometer from within a sample vessel (test tube, multiwell plate or similar container). At the desired time for analysis, sample is aspired from this vessel and is transported through tubing to a ?ow cell or nozzle. At the ?ow cell, the sample is introduced to the center of a faster ?owing carrier ?uid and is in turn presented to one or more light sources for excitation. Light scatter and/or ?uorescence are captured, spectrally ?ltered and directed to appropriate photo detectors for conversion to electrical signals. Electronic circuitry is used to process these signals for analysis, classi?cation, sorting and data storage. For sorters, an individual particle is traced through the instrument as it breaks free from the continuous jet into a charged droplet for electrostatic de?ection. The droplet passes through an electric ?eld and is ultimately captured in a suitable container for further processing or study, or is disposed of as waste.

Application in Seed Quality Analysis

1. Identification of Discrimination in the Quality of Seed Lots obtained from different Growers

Not often, however, do seed growers have access to flow cytometers; they mostly rely on their experience to visually estimate seed maturity. Hence, the seed industry receives partly processed (dried and cleaned) seed lots of different quality from different growers. Can the quality of these lots be determined by flow cytometry? Studies on sugarbeet seeds show that it can be. By analysing fresh and dried seeds collected at five different times during maturation drying, starting from the stage when the embryo was already fully developed. It was found that after drying there are significant differences in the proportions of the 4C embryo nuclei, and both 3C and 6C endosperm nuclei, between seeds of different maturities. In seeds dried at an early stage of maturation, the proportion of the 2C embryo nuclei is higher than in those harvested after completing maturation drying. Probably because of lower tolerance of the nuclei with a higher DNA content to rapid water loss, in early-harvested seeds these nuclei were not able to withstand desiccation, whereas slow drying on the mother plant (to a water content below 30 per cent) prevented damage during after-harvest drying. Consequently, seeds with a higher 4C/2C ratio (fully mature) exhibited a higher vigour and germination capacity. Therefore based on the 4C/2C ratio discrimination in the quality of seed lots obtained from different growers can be identified.

2. Establishment of Seed Harvest Time

Observations on cell cycle activity allow determination of the time of transition from germination to seedling growth, which is when seed desiccation tolerance starts to be lost. This information is useful for establishing an appropriate seed harvest time, especially when there is rain during the maturation/ripening period that can induce pre-harvest sprouting.

3. Information on Seed Dormancy

Flow cytometric analysis can also provide information on seed dormancy. In imbibed dormant tomato seeds, nuclear DNA replication is blocked (the proportion of the 4C nuclei does not increase above that observed in dry seeds). Breakage of dormancy induces an increase in the number of 4C nuclei in the radicle tip, which precedes its protrusion. Similarly, the removal of dormancy from cherry (*Prunus avium*) and Norway maple (*Acer platanoides*) seeds by stratification coincides with a change in nuclear DNA content.

4. Monitoring Seed Priming

Seed quality can be improved by processing and pre-sowing treatments such as priming. These treatments, which are based upon controlled hydration of the seeds (using water or non-osmotic and osmotic solutions), promote germinative metabolism; since this promotion is retained after subsequent dehydration, it leads to rapid and uniform seedling emergence in the field. One of the problems that producers face when applying priming to seeds is that it has to be finished before completion of germination, otherwise there is a decrease in germinability caused by a loss of desiccation tolerance following radicle protrusion. Additionally, there is variation among seed lots, even of the same species, in their response to priming, and thus conditions of the treatment have to be optimized individually for each lot. Depending on the species, laboratory germination tests, which are routinely applied to assess seed quality control, take several to a dozen or more days and therefore slow down the final processing of the seeds, exposing the seed industry to economic losses. Molecular markers of germination that allow an immediate check of priming advancement, including DNA replication, could reduce these losses.

Changes in nuclear replication stages upon priming have been studied by flowcytometry, mostly on pepper, tomato and sugarbeet, species producing orthodox seeds. The rate of DNA replication during priming depends on the seed moisture content as well as on temperature and oxygen availability. In most osmoprimed pepper seed lots there is advancement of embryonic root tip nuclei into S and G_2 phases of the cell cycle, as measured by an increase in the percentage of nuclei with 4C DNA content, although DNA replication is retarded compared to that of seeds imbibed in water. The appearance of 4C nuclei in pepper radicle tips is usually first noticeable after about a week of incubation in osmotic solution, and after 2 weeks they exceed 40 per cent. Moreover, the lag period necessary for DNA repair before its replication during subsequent introduction of primed seeds to germination conditions, was reduced to as little as 12h (from 36 h for untreated seeds similarly imbibed in water), when the osmopriming was applied for 14 or 21d. Significant correlations were found between the frequency of priming-induced nuclear replication and the improvement of pepper seed vigour, as measured by the reduction in mean germination time. Thus, the 4C/2C ratio can be used to predict seedling performance of this species.

Flow cytometry can also be used for establishing germination advancement when, after priming, the seeds are to be stored.

5. Estimation of Seed Lot Purity

Seed plants are often exposed to the risk of pollination by their wild relatives. The contaminations can be distinguished by estimating their ploidy. During hybrid seed production in the field there is also the danger of including some seeds with the tetraploid male component. The presence of seeds of other than triploid ploidy can be easily detected by flow cytometry at any developmental stage, also directly in commercial lots of seeds partly processed by growers, using any part of the true seed removed from the pericarp. This kind of analysis is already routinely used by seed producers, who control contamination with undesirable ploidies in all seed lots obtained from different growers by eliminating those with a considerable proportion of diploid and/or tetraploid seeds.

Water Activity Meter

Water in seeds is a complex system and is only partially understood. Even though much is not known about water-plant substances relationships, it is known that water is associated with the seed system in several patterns. In some cases, it is actually part of the chemical structure of other molecules of the seed tissue, held by hydrogen bonding (vectorized polar bonds), and does not exist as discrete water molecules. In other cases, water is held as discrete molecules in bonding interactions with seed tissue molecules, though the arrangement and stability of this type of water is highly variable. These interactions may extend into the surrounding liquid, forming gradient patterns of structure in a dynamic state of turnover. Hence water in seeds have been characterized into three phases or zones depending on the way it is held by the plant substances.

Bound water is tightly held to ionic groups such as amino or carboxyl groups and exists as a monolayer around macromolecules of the seed. Adsorbed water is considered to exist in multilayers, loosely held by bonding to hydroxyl and amide groups above the monolayer of bound water. Free water is considered as capillary or solution water held only by capillary forces to the seed tissues. However, water may always in association with other systems, and at which point water is bound and free is difficult to ascertain.

Basics of Water Activity

Traditionally, discussions about water in seeds focus on moisture or water content, which is a quantitative or volumetric analysis that determines the total amount of water present. The limitations of water content measurement as an indicator of safety and quality are attributed to differences in the intensity which water associates with other components in the seed. The water content of a safe seed varies from seed to seed. One safe, stable seed might contain 15 per cent water while another containing just 8 per cent water is susceptible to microbial growth. Although the wetter seed contains proportionally more water, its water is chemically bound by other components, making it unavailable to microbes. Using only water content values, it's impossible to know how "available" the water in the seed is to support microbial growth or influence seed quality.

Another more important type of water analysis is water activity (aw). Water activity describes the energy status or escaping tendency of the water in a sample. It indicates how tightly water is "bound," structurally or chemically, in seeds. Both the water content and the water activity of a sample must be specified to fully describe its water status. However, water activity is the property most relevant for quality and safety issues. Water activity is closely related to the partial specific Gibbs free energy of the system. Thus, water activity is a thermodynamic concept and has requirements for measurements. These requirements are that the system be in equilibrium, the temperature defined, and a standard state specified. Pure water is taken as the reference or standard state from which the energy status of water in seed is measured. The Gibbs free energy of free water is zero; thus, the water activity is 1.0.

Water activity is the ratio of the vapor pressure of water in a material (p) to the vapor pressure of pure water (po) at the same temperature. Relative humidity of air is the ratio of the vapor pressure of air to its saturation vapor pressure. When vapor and temperature equilibrium are obtained, the water activity of the sample is equal to the relative humidity of air surrounding the sample in a sealed measurement chamber. Multiplication of water activity by 100 gives the equilibrium relative humidity (ERH) in per cent.

aw = p/po = ERH (per cent)/100

As described by the above equation, water activity is a ratio of vapour pressures and thus has no units. It ranges from 0.0aw (bone dry) to 1.0 aw (pure water).

Instrumentation

AquaLab Dew Point Water Activity Meter 4TE (Figure 20.2) is a recent model widely used for water activity measurement.

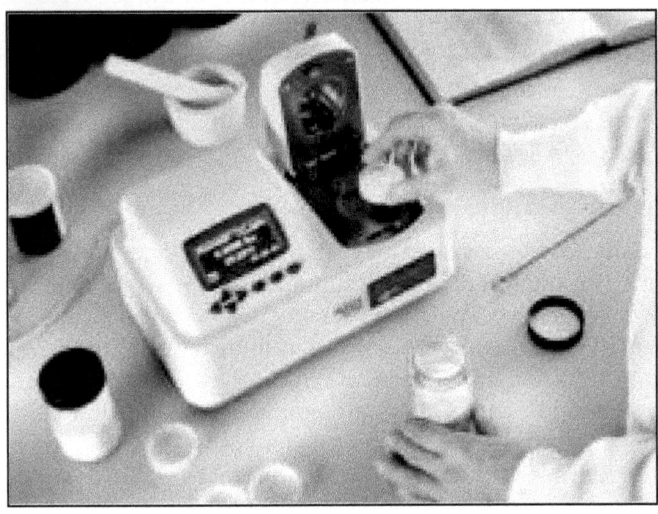

Figure 20.2: AquaLab Dew Point Water Activity Meter 4TE.

Water activity is measured by equilibrating the liquid phase water in the sample with the vapor phase water in the headspace of a closed chamber and measuring the relative humidity of the headspace. Methods for water activity determinations are detailed in the Official Methods of Analysis of AOAC International (1995).

Two different types of water activity instruments are commercially available. One uses chilled mirror dewpoint technology while the other utilizes relative humidity sensors that change electrical resistance or capacitance. AquaLab Water Activity Meter 4TE works on the principle of chilled mirror dewpoint technology.

In a chilled mirror dewpoint system, a sample is placed in a sample cup which is sealed against a sensor block. Inside the sensor block is a dewpoint sensor, an infrared thermometer, and a fan. The dewpoint sensor measures the dewpoint temperature of the air, and the infrared thermometer measures the sample temperature. From these measurements the relative humidity of the headspace is computed as the ratio of dewpoint temperature saturation vapor pressure to saturation vapor pressure at the sample temperature. When the water activity of the sample and the relative humidity of the air are in equilibrium, the measurement of the headspace humidity gives the water activity of the sample. The fan is to speed equilibrium and to control the boundary layer conductance of the dewpoint sensor.

The major advantages of the chilled mirror dewpoint method are speed and accuracy. Chilled mirror instruments make accurate (± 0.003aw) measurements in less than 5 minutes. Since the measurement is based on temperature determination, calibration is unnecessary, but running a standard salt solution checks proper functioning of the instrument. If there is a problem, the mirror is easily accessible and can be cleaned in a few minutes. For some applications, fast readings allow manufacturers to perform at-line monitoring of a product's water activity.

Application of Water Activity in Maintaining Quality during Seed Storage

Prior to germination, seeds need to be stored at water activities that will keep them stable by minimizing enzymatic activity that could reduce their viability but also avoid complete desiccation that will also result in loss of viability. Although the range of viable water activity values is somewhat wide, in moisture terms, a change of as little as 1 per cent can damage germination rates.

Another problem encountered during storage is fungal growth. Knowing the moisture content of your seeds doesn't tell you anything about what can grow in them during storage. Water activity values corresponding with potentially harmful bacteria are well-defined (Table 20.1). Water activity better predicts the growth of microorganisms because microorganisms can only use "available" water, which differs considerably depending on the solute. On average, ions bind the most water, whereas polymers bind the least water; sugars and peptides fall into an intermediate position. At the same molecular concentration, salt lowers the water activity more than sugar.

Table 20.1: Common Spoilage Organisms and their a_w Limits for Growth

Sl.No.	*Microbial Group*	*Example*	a_w
1.	Normal bacteria	*Salmonella* sp., *Clostridium botulinum*	0.91
2.	Normal yeast	*Torulopsis* sp.	0.88
3.	Normal molds	*Aspergillus flavus*	0.80
4.	Halophilic bacteria	*Wallemia sebi*	0.75
5.	Xerophilic molds	*Aspergillus echinulates*	0.65
6.	Osmophilic yeast	*Saccharomyces bisporus*	0.60

In the above table most of microbial activity is ceased below water activity of 0.60. The following example of corn and mustard seed clearly depict the importance of water activity over water content measurement and also microbial interaction with water activity.

The three vertical lines in Figure 20.3, at water activities of 0.60, 0.85, and 0.94 represent limits below which, respectively, no spoilage occurs, products are defined as shelf stable, and toxin production by botulism is stopped.

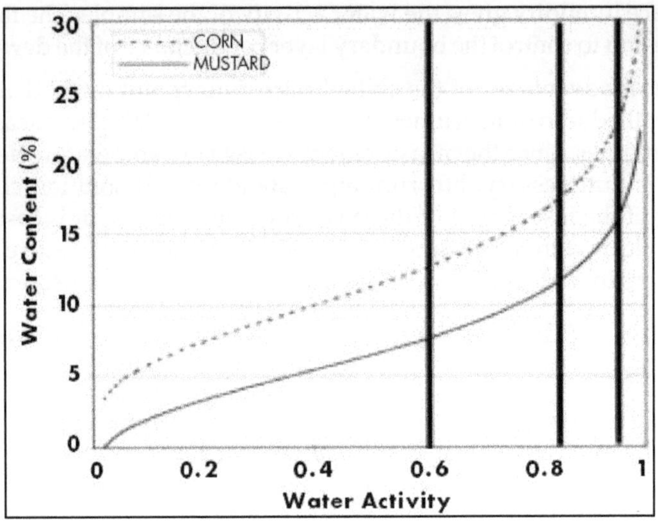

Figure 20.3: Isotherms of Corn and Mustard Seeds.

From the figure it is clear that corn can be stored without spoilage if its water content is below 12.5 per cent, and mustard is similarly safe if its water content is below 7.5 per cent. In fact, these are values that have been used for years to determine whether these products are safe to store. But the water content below which spoilage fungi are inhibited is different for each different product. This is because it is the water activity, not the water content that determines whether these organisms can carry on their physiological functions. When the water activity is below 0.6, water is so tightly bound by the product that it becomes unavailable to even the most xerophytic fungi.

Water content is irrelevant in this application because the microbes respond to the availability, not the amount of water. Thus, even though it is common practice to measure water content for this application, water activity is a much better measure. Water activity is not product specific, and is fundamentally related to the process we are interested in.

In general, water activity is the right measurement when microbial processes, including seed spoilage, are of concern. Also, moisture migration, physical properties and caking are best addressed using water activity.

Keyence Digital Microscope

One of the major objectives in seed testing laboratories is the positive identification of seeds in purity tests and determination of other seeds. The absence or presence of certain seeds such as weeds can determine whether see lots may be imported into certain countries or not if national regulations are restrictive.

Next to specialized literature, seed collections and herbaria are traditionally used for the precise identification of species. The major importance of seed collections is to allow the direct comparison of the unidentified seed with seeds from certified reference material.

To establish such a reference seed collection is a huge challenge, requiring first the collection and identification of plants, then the harvesting of the seeds, possibly from different plants at different stages of development, in order to show the variability. In further steps, the seeds must be dried carefully, processed (usually by hand), labelled, stored and protected from insects.

Nowadays, seed testing laboratories have the problem that seed collections are either not available for purchase, or only at high cost, and laboratories may not be able to afford a large number of species.

Technical progress has opened up new possibilities using photograph and computer programs. They permit three dimensional illustrations of seeds with a vivid view of the structure and surface, which even allows the identification of seeds by morphological characteristics. This creates the possibility of establishing a digital seed collection.

Advantage of Digital Seed Collection

The major advantage of such a digital seed collection is that the seed characteristics needed for identification are all visible on the screen, without any further tools such as magnifying glasses or binocular microscopes being necessary. The quick use and availability of the data base at different locations in a lab are as convenient as the security in working with reference seeds, since there is no danger of returning seeds to the wrong box.

Maintenance of the seed collection is also easier. The shape and colour of the photos remains bright, and the visible seed characteristics remain unchanged, in contrast to real seeds, in simple compared to real seeds, in which the process of ageing leads to shrivelling, loss of colour saturation or even change of colour. Also data management is simple compared to a real seed collection, and such a photo collection may be continually expanded.

Instrumentation

The name "Keyence" stands for "Key of science" a technique developed in 1974 in Osaka, Japan, originally for application in industrial sensory and automation technology.

The Keyence digital microscope system consists of a tripod, a telephoto zoom lens, a personal computer with a high resolution screen and the necessary software to make and create three dimensional images (Figure 20.4). The system includes sources of cold light. The use of the Keyence is quite simple and allows enlargements from 20 - 200× or 500 - 5000× in one operation.

Seed identification is usually carried out using characteristics such as size, shape, colour and especially structures of the seed surfaces, which in some cases may even be species specific.

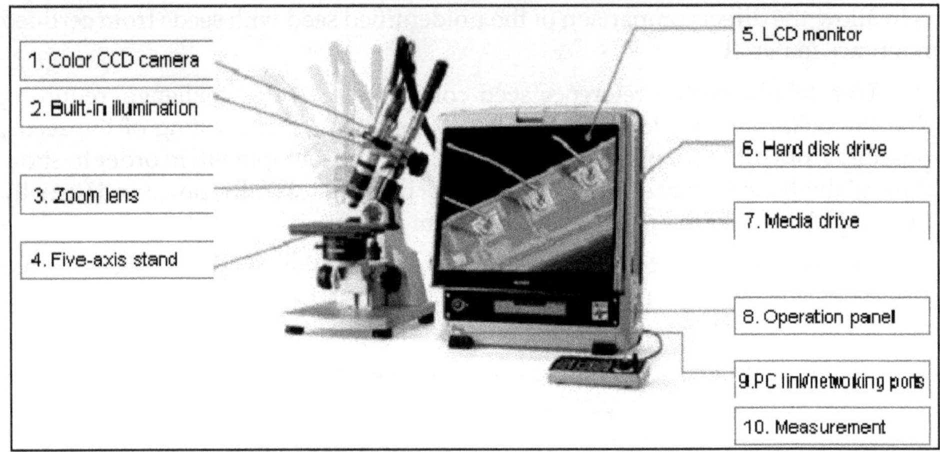

Figure 20.4: Components of Keyence Digital Microscope.

Working with the Keyence System

For best results with the Keyence microscope, the settings must be optimized to create the best conditions for displaying the morphological features necessary for species identification. For instance, lighting may be required which provides the right combination of light and shade to enhance the three dimensional appearance, but which at the same time avoids bright reflecting spots on the shiny surfaces of seeds such as beans, millet or sorghum. The software then offers various interactive modes which are suitable for individual seed structures, as shown in the following examples.

Q2 Analyzer

Q2 analyzer works based on the principle of oxygen consumption of individual seeds of a seed lot. It measures the oxygen consumption each seed precisely with at most rapidity. It allows a fast but accurate measurement of the germination level of a seed lot. Depending on the species, the estimated time needed will be between 10 and 48 hours. You will easily determine dead, dormant or actively germinating seeds. Although it cannot give you any specific details on seedling abnormalities, the Q2 can give quick and accurate indications of the vigor and the homogeneity of a seed lot. It can give us insight in speed and uniformity of germination of a specific seed-lot. But the only demerit is that the Q2 tests are destructive.

Applications

- ✰ Separation of vigorous from non-vigorous seeds for breeding purposes
- ✰ Measuring the vigour and the speed of germination of a seed lot
- ✰ Detection of dormant seeds versus dead seeds
- ✰ Parameter calculations in priming processes
- ✰ Control and detection of fungal infections
- ✰ Insights in the homogeneity of a seed lot

- Fast prediction of the germination
- Control of oxygen sensitivity
- Control of seed metabolism.

Chapter 21

Farmers Producing Company: An Innovative Institutional Intervention for Agribusiness

R. Sendilkumar[1] and G. Naveenkumar[2]

[1]*Professor (Agrl. Extension),*
Kerala Agricultural University, Thrissur, Kerala
[2]*PG Student (Agrl. Extension),*
Department of Agricultural Extension, College of Horticulture,
Kerala Agricultural University, Thrissur, Kerala

Prologue

The growth rate of Indian agriculture over the last decade has been declining and it is a matter of serious concern for all those stakeholders associated with agriculture. Several institutional intervention models were tried in India to integrate farmers, inclusive of small farmers with the value chain perspective. The most promising model is the Producer's Cooperatives which enable farmers to organize themselves as collectives. Except very few like AMUL, the cooperative experience in India has not been a pleasant one as it has been infected by several inadequacies. Cooperation is a Government triggered programme rather than people movement in India. Cooperatives have largely been State promoted, with a focus on welfare rather than business on commercial lines (Prabhakar *et al.*, 2012). Amidst such deficiencies and inadequacies in Cooperative system, there was an attempt in 2002 to strengthen the Cooperative movement with the amendment of Companies Act,

1956 under the Chairmanship of Professor Dr. Y. K. Alagh, which paved the way for incorporation of "Producer Company" (PC) (Singh, 2008). Producer Company means a body corporate having objects or activities specified in section 581B and registered as Producer Company under this Act. Producer Company is the hybrid between a Private limited company and a Cooperative Society (Venkattakumar and Sontakki, 2012).

In a Producer Company, only persons engaged in an activity connected with or related to primary produce can participate in the ownership. The basic purpose of the FPC is to group small farmers for backward linkages i.e, for inputs like seeds, fertilizers, credit, insurance, knowledge and extension services, and for forward linkages such as collective marketing, processing and market led agriculture production (Sebastian, 2013).

National Sample Survey Organization (NSSO) reported that given the choice, 40 per cent of the farmers, wishes to leave agriculture, and because of non-remunerative price that has been realized by the primary producers has been one among the prime reasons. There has also been no surplus produce for value addition due to low productivity influenced by poor knowledge base towards production technology, access to credit, input, market and obviously the below-par adoption behavior (Murray, 2009). To make the famers to start their own corporate entity and to become a business man, the Government of India had framed the concept Farmer Producer Company.

What is a Producer Company?

"Producer Company" means a body corporate having objects or activities specified in section 581B and registered as Producer Company under this Act (Companies Amendment Act, 2002). A Producer Company is a democratically controlled enterprise owned by the community and managed by Professionals.

Objects of Section 581B deals with

- Production, harvesting, procurement, grading, pooling, handling, marketing, selling, export of primary produce of the members or import of goods or services for their benefit.
- Processing including preserving, drying, distilling, brewing, canning and packaging of produce of its members.
- Manufacture, sale or supply of machinery, equipment or consumables mainly to its members.
- Providing education on the mutual assistance principles to its members and others;
- Rendering technical services, consultancy services, training, research and development and all other activities for the promotion of the interests of its members.

Private Limited Company Vs Cooperative Society Vs Producer Company

Private Limited Company	*Cooperative Society*	*Producer Company*
Companies act	Cooperative Societies act	Companies act
Minimum two members and maximum 50	Minimum 50 members	Minimum 10 primary producers (Or) two producer Institutions
Membership opens to all	Open only to individuals and cooperatives	Only those who participate in the activity
Shares are tradable and transferable	Shares are not tradable	Not tradable but transferable
Voting rights based on number of equity shares held	One person, one vote, but Government and RCS holds veto powers	One person one vote
Can have two types of shares Equity and preference	Only equity type of shares	Only equity type of shares
Conversion of private limited to PC is not possible	Conversion of cooperatives to PC is possible	Conversion to private limited is not possible

Source: Compiled by the author.

Who can form Producer Company??

- ☆ Any ten or more persons engaged in any activity connected with primary produce, or
- ☆ Any two or more producer institutions or Companies, or
- ☆ A combination of ten or more individuals and producer institutions.

Characteristics/Features of Producer Company

- ☆ The Producer Company should be registered under the Companies Act
- ☆ The minimum numbers of members have to be ten
- ☆ Producer Company is formed with limited liabilities and limited only by share capital.
- ☆ The liability of the members is limited to the unpaid amount of the shares held by them.
- ☆ The voting rights shall be based on a single vote for every member, irrespective of his shareholding or patronage.
- ☆ The member would be the primary producers. The PC will have a management team to conduct day to day operations and will be governed by a board of members.
- ☆ The PC would have a Board elected from among the members. The Board may co-opt one or more expert directors or an additional director not exceeding one- fifth of the total number of directors.

- ☆ The surplus arising out of the operations of the Producer Company shall be distributed in an equitable manner by providing for the development of the business.
- ☆ It shall never become a public (or deemed public) limited company.
- ☆ Members' equity cannot be publicly traded but be only transferred.

Advantages of Producer Company

- ☆ Training members on good agricultural practices based farming system approach and low-cost and environmental friendly inputs
- ☆ Arrange finance, capital and working capital
- ☆ Advising crops to be grown
- ☆ Preparing and providing inputs at low-cost
- ☆ Procuring produce at a price committed at the time of sowing/planning
- ☆ Procuring firm orders from market and Government programs
- ☆ Adding value to the farm produces locally and thus adds to profitability of the Company
- ☆ Market value-added products at maximum profits and thus generate funds for salaries of its employees, generate reserves of cash and profits for the company.

Initiators of Producer Company

Initiator could be a person or a group of persons who takes the responsibility to initiate and establish a producer company. Further, initiator could also be one of the promoters of the company. The type of initiators includes

- ☆ Person (not necessarily a primary producer),
- ☆ Interested group of persons willing to contribute their time and resources to promote a producers company,
- ☆ Any NGO working with the primary producers group,
- ☆ The existing Multi-State Cooperative Societies or Federation of Co-operative,
- ☆ Any Community Organization, whose members are interested in the concept,
- ☆ Any Government organization or department willing to promote a producer company.

The role of the initiator slowly transform to facilitators they undergone the process of empowerment.

Stages in the Evolution of FPC

There are three stages in the evolution of Producer Companies in India (Figure 21.1).

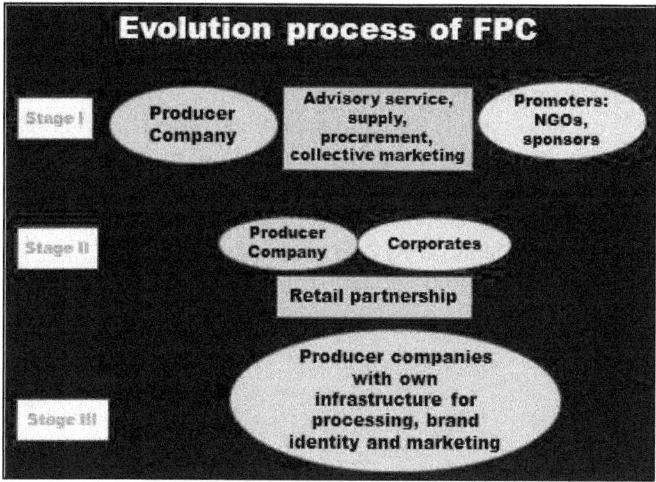

Figure 21.1: Stages in the Evolution of Producer Companies in India (*Source*: Murray, 2009).

1. **First stage of evolution of producer companies**: In general, most of the initiatives on producer companies are start-ups and promoted by NGOs/ development agencies/sponsoring organizations. Mostly they do the function of providing technical services, input providers and facilitator for collective marketing.
2. **Second stage of evolution of producer companies:** Corporate comes together with farmers to share prosperity through retailer partnerships. *e.g.* FPC promoted my Fab India.
3. **Third Stage of evolution of Producer Companies**: Here, FPC developing their own identity, brands, established the stable supply-chain and command over the market pie.

Steps in Incorporation of FPC

There are four main stages in the establishment of a Producer Company as described below:

1. **Initiation:** This includes the process formation, in which external agency mobilizes the farmer for the establishment and to become shareholders for the company.
2. **Registration of name:** It has to undergo three sub stages *viz.*, a) Digital Signature Certificate, b). Director Identification Number (DIN), c) Naming of a Producer Company
3. **Filling legal documents:** It involves a) Memorandum of Association, b) Articles of association,
4. **Certificate of incorporation**

Typology of Producer Companies in India

There are two directions of evolution in FPCs in India i.e, one direction of evolution is from an inward- to an outward-oriented type of producer company (change from type A to type B, or from type C to type D). The other direction of evolution is from being a FPC, focusing on trade with agricultural inputs to one that focuses on marketing its members? produce (change within type C or type D). Type A and type B are promoted by NGOs/government organizations and whereas type C and type D are promoted by corporate (Trebbin, 2014).

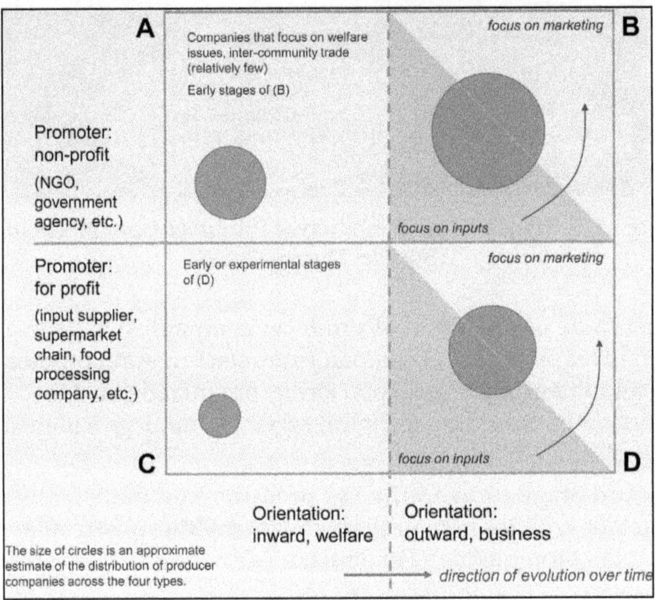

Figure 21.2: Typology of Producer Companies in India. (*Source*: Trebbin, 2014).

Financial Support to FPC

Financial support is available from many sources. Some of them are

☆ Producers Organization Development Fund (PODF)

The fund has been created in NABARD and it will be used to support Producers Organizations across three level, *viz.* credit support, capacity building and market linkage. The objective of the fund is to meet end to end requirements of Producers Organization as well as to ensure their sustainability and economic viability. (https://www.nabard.org/).

☆ Farmers' Technology Transfer Fund (FTTF)

The main objective of this fund is to support the activities of farmers should result in increase in farmer's income through increased production, productivity, reduction in costs, demand led diversification, market participation, value addition *etc.* (https://www.**nabard**.org/).

☆ Indian Grameen Services (IGS)

Indian Grameen Services (IGS) has initiated an innovative loan produc,t where they provide loan to a FPO on the credibility of the promoting organization and FPO itself, without any collateral.

☆ Friends of Women World Banking [FWWB]

Friends of Women World Banking [FWWB] have a corpus of ₹1.5 crores and provide loans to FPOs on the basis of their business Plan without any collateral security.

☆ Department for International Development (DFID)

DFID supports India's poorest states like Madhya Pradesh, Odisha, Bihar and West Bengal, there it works with an NGO called Action for Social Advancement (ASA) for the promotion of Farmer Producer Companies for the up liftment of the rural poor farmers in that area.

☆ World Bank

World Bank provides its support to FPCs through a programme called Madhya Pradesh District Poverty Initiative Project (MPDPIP) and routed through an NGO Action for Social Advancement (ASA).

Institutional Support

Various institutions have been giving direct and indirect support for the promotion of FPC. Some of them are

- ☆ Small Farmers Agribusiness Consortium (SFAC)
- ☆ Coconut Development Board (CDB)
- ☆ Indian Society of Agribusiness Professionals (ISAP)

Various NGOs like

- ☆ PRADAN- Professional Assistance for Development Action
- ☆ ASA- Action for Social Advancement
- ☆ ESAF- Evangelical Social Action Forum

Challenges

Though there are lots of financial and promotional supports, the growth is not adequate and faces lot of practical difficulties to face various challenges. The challenges reported by the researchers are:

☆ Lack of Awareness

There is very little awareness about Farmers Producer Companies [FPCs] amongst the farmers and institutions deal with, the corporate sector, the input suppliers, the commercial banks, even the district level officials and agriculture department officials.(NRAA, 2009).

☆ Capital Formation

Working capital assessment as well as availing the capital for running Producer Company from financial institutions whether from commercial banks or even from NABARD is a herculean task. FPCs/Cooperatives, do not have the assets that is required as collateral (Pustovoitova, 2011).

☆ Capacity Building

The concept of farmers, as producers and business perspective is a newly emerged one and challenging. Therefore, Capacity building of grass root level functional coordinators (local resource persons), members of the Board of Directors of the producer company is extremely important for viable. (Puzniak and Cegys, 2011).

☆ Market Linkages and Value Addition

Producers lack information regarding demand and supply conditions of the produce. The villages not established either warehousing or cold storage facilities. The price fluctuations and the lack of market linkage pose huge challenges before FPC. Value addition will require huge investment Capital.

☆ Input Supply

There is short supply of inputs in the villages and artificial shortages are created during the peak season. Even manufacturers are involved in this to increase their profit. FPC/Cooperatives don't have sufficient capital to source the raw materials from various agencies.

☆ Need for External Catalysts

Farmers Producer Company formation requires an external agent to mobilize the farmers first in a common interest group and then federate a number of such groups in to a Company. Continuous hand-holding is required to build capacities.

Strategies to Overcome the Challenges

1. Education and awareness regarding FPOs has to be raised amongst the general body members of the FPOs through well built capacity building programme.
2. Promoting FPCs through alternative routes such as SHG route, Corporate route, NGO/Government promoted route:
   - ☆ **SHG route:** Development department, some non-governmental organizations like PRADAN and SERP build from a base of women's SHG that federate into women based FPCs.
   - ☆ **Corporate route:** The private sector agri-input company may provide grant finance for promotion and running of the FPO. This is in addition to seeds and technical support. The input supplier company's interest would be promotion and sale of their own products to the exclusion of other manufacturer's products.

- ☆ **NGO/Government promoted route:** This consists of the promoting organization mobilizing farmers into Farmers Interest Groups [FIGs] first and then federating them into FPCs. The FIGs and FPCs deal with a variety of produce raised by the farmers and is not produce–specific. It is widely agreed that an external agency is required to guide and assist the FPOs (Nair, 2013).

3. Promote FPO in engaging input supply business. This type of initiative and entry will prompt other retailers dealing in agri-inputs to lower/ rationalize the price they charge from farmers.
4. FPOs should set up agri-clinics with skilled manpower that can provide professional services to farmers.
5. FPO can encourage starting custom hiring services of agricultural machinery especially to small and marginal farmers who do not have the capital to purchase big tools and equipment.
6. Warehouse construction may be given special consideration if built by FPOs. Or management of warehouses constructed with public funds may be given to FPOs.
7. Market and other agriculture related information can be passed on to members through SMS or Community radio service. Farmers can also be provided with toll free number which they can use to access information according to their need.
8. Knowledge building to understand the business viability of FPOs is needed. Tracking loan disbursements to FPOs and repayment percentages by them should be taken up by a nodal agency at the State level. This is to create a database which will help establish the credibility of the FPOs (Puzniak and Cegys, 2011).

Case Study

Palakkad Coconut Producers Company Limited. (PCPCL) venturing in Neeera Production, Processing and Marketing

Palakkad Coconut Producer Company limited is an initiative of the coconut farmers of Palakkad district formed in June 2013. It is the apex body of three tier structure that includes Coconut Producer Societies (CPS) and Coconut Producer Federations (CPF) (Figure 21.3). It is the 2nd producer company registered in Kerala upon facilitation by the Coconut Development Board. The main objectives of the company are to build a prosperous and sustainable coconut sector by forming a farmer owned Producer Company that enables the farmers to enhance productivity through efficient cost effective and sustainable resource use.

PCPCL is currently constituted of 413 CPSs which are constituted in to 23 federations of CPSs. Of which 17 Federations of Coconut Producer Societies are participated to form the company and leading the initiatives of coconut procurement, copra dryer and tender coconut stalls. Substantial part of these efforts is being financed through contributions from farmers. It also carried out the business of producing, manufacturing, marketing, importing, exporting, developing, and dealing coconut based products.

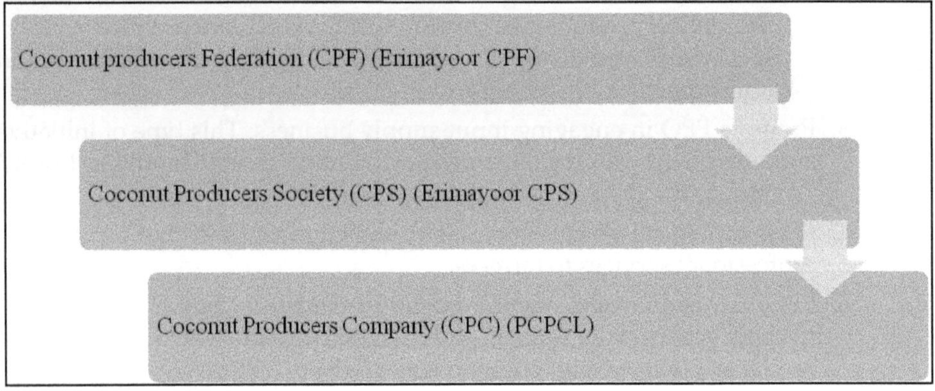

Figure 21.3: THREE Tier Structure of Company.

Neera Production, Processing and Marketing

Neera production, processing and packaging undertaken by the Federations of CPS registered with CDB (Figure 21.4). License for Neera production was issued to Federations of CPS registered with CDB. Neera collected by the Federations through the member CPS brought to a primary processing centre which is located at Mudualamada Palkakkad district. Further, processing and logistics distribution were carried out at PCPCL.

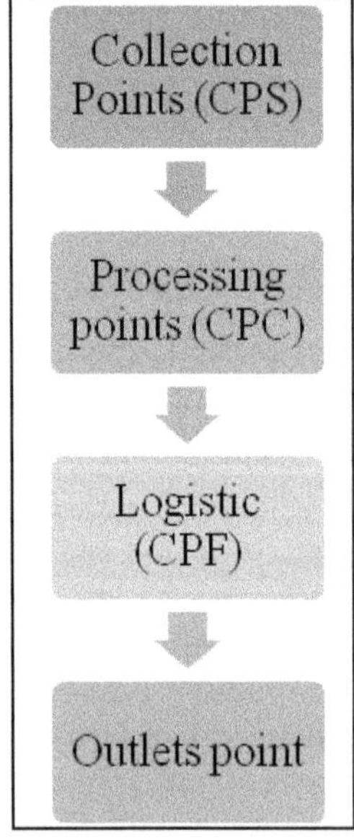

Figure 21.4: Neera Collection, Processing, Logistics and Marketing of PCPCL.

Muthalamada Federation of CPS obtained the license to manufacture, process and tapping of neera. At present, 607 spadix from 461 palms, owned by 27 farmers have been tapped in Muthalamada. The spadixes tapped at various stage of growth. Federation tapped around an average of 350 litres of neera per day. The process of tapping carried out by specially trained neera technicians, who collect the sap from the spathe of healthy coconut palms from selected groves notified for this purpose. After tapping neera from the mature spadix twice a day, the harvested neera shifted from the palm top ice boxes to the insulated boxes. Refrigerated vehicles have been used to transport the harvested neera to the cold storage unit for processing and distribution. The harvested neera maintained in the temperature range of 4 to 10 degree celcious throughout extraction, processing and transporting stages. Quality ensured at each stage by following standard procedures and retaining original flavor, aroma and taste. Good management practices are being followed to ensure quality and quantity from the place of production to consumption point. The establishing and operating facilities of cold chain for harvesting, processing, and storing of neera from 1500 palms in Muthalamada would cost around ₹80 lakhs.

Neera is made available to the consumers through vending machines placed in 17 important places of town and tourist spot. The cost of Neera @ of 30 per cup measuring 200 ml. By- products like Palm sugar and Jaggery also produced from the waste pulp and packed in 100, 250, 500, 1000 gram sachets. The packed Neera shelf life is 3-4 months in room temperature and over a year in refrigerated conditions. However, PCPL sells neera only through their out lets distributed through well designed logistic services.

The Company procuring coconuts tender coconuts and Neera (un-fermented coconut sap), palm sugar and other produce from the shareholders and resells the raw or value added produce to various governmental and business agencies and through its own outlets. The produce collected from each farmer was proportionate to his/her number of shares.

References

Murray, E. V. 2009. Producer company model: Opportunities for bank finances. CAB calling.

NABARD (National Bank for Agriculture and Rural Development). 2013. NABARD home page (online). Available: htpp: //www. Nabard.org. [07 Aug. 2013].

Nair, D. 2013 Producer companies the new outlook in the global world. *Indian Coconut J.* 56 (5): 15-16.

NRAA. 2009."Perspectives and problems of primary Producers Companies - Case study of Indian Organic Farmers Producer Company Limited, Kochi, Kerala"; National Rain fed Area Authority, New Delhi, India. 30p.

Prabhakar, I. T., Manjunatha, B. L., Nithyashree, M. L., and Hajong, D. 2012. Farmers Producer Company (FPC) - an innovative farmers' institution. *Environ. and* Ecol. **30**(2): 427-430.

Puzniak, M. and Cegys, P. 2011. Producer Companies and their partnered institutions, owned by small and marginal producers in India, M.Sc. thesis, Lund University Master'sProgramme in Environmental Studies and Sustainability Science (LUMES), Sweden, 63p.

Pustovoitova, N. 2011. Producer Company as an institutional option for small farmers in India. M.Sc. thesis, Lund University Master?s Programme in Environmental Studies and Sustainability Science (LUMES), Sweden, 71p.

Sebastian, K. S. 2013. Producer Companies for prosperity. *Indian Coconut J.* **56**(5): 4-7.

Singh, S. 2008. "Producer companies as New Generation Co-operatives". *Econ. and Polit. Wkly.* **43**(20): 22-24.

Trebbin, A. 2014. Linking small farmers to modern retail through producer organizations – experiences with producer companies in India. Food Policy. **45**: 35–44.

Venkattakumar, R. and Sontakki, B. S. 2012. Producer companies in India-experiences and implications. *Indian Res. J. of Ext. Educ.* **1**: 154-160.www. asabhopal.in/https: //www.nabard.org/.

Chapter 22

Empowerment Dynamics of SHG, Enterprising in Food Supplements

R. Sendilkumar[1] and A.K. Devi[2]

[1]Professor (Agrl. Extension),
Kerala Agricultural University,
Vellanikkara, Thrissur – 680 656, Kerala
[2]Graduate Scholar,
Kerala Agricultural University, CCBM,
Vellanikkara, Thrissur – 680 656, Kerala

ABSTRACT

Women empowerment process is one, where women find time and space of their own and begin to re-examine their lives critically and collectively approaching problems in different ways. The Central as well as State Government promotes SHG to start income generation activities. As a result many SHGs have been engaged in enterprising. By doing so, it is also expected that they should have got empowered in many dimensions inorder to attain sustainability. Keeping this as core concept, SHGs engaged in production of food supplements in Thazhavagrama Panchayat of Kollam district of Kerala were selected. Accordingly two SHGs were considered for the study with total sample size of 50 women participants who have concentrated in the production of food supplements. Inteview schedule and focussed group discussion were employed to collect the data. The empowerment concept was studied in five dimensions. An overall 32 per cent increase in the empowerment dynamics index (EDI) was noticed as cumulative effect of all the components of empowerment. An appreciable increase observed for all the dimensions of empowerment viz., Psychological component (43 per cent), Social (41 per cent), Economic and Entrepreneurial (37 per cent), Political and Creative ability (24 per cent), and Knowledge component (8 per cent) in order. In nutshell, the empowerment dynamics of women participants of the SHGs have reportedly increased after joining the SHGs. Hence the study gave a strong message that women SHGs assumed an important place in the overall growth of women who lived in rural areas and also enhance the production of food supplements in the grass root level ensuring the nutritional security to vulnerable sections.

Keywords: *Empowerment dynamics, SHG.*

Introduction

Any attempt to improve the status of women should start with 'empowerment'. Empowerment implies the creation of an enabling environment, where individual can fully use their capabilities to accomplish to take charge of their lives. Even though women are greatly participated in almost all sectors of economy, their activities as producers are not adequately recognized at the national level. In the Eleventh Five Year Plan (2007-2012), for the first time, women were recognized as just equal citizens, but as agents of economic and social growth (Plan report 2007-2011). Accordingly all the policy initiatives and administrative efforts triggered for women development programme and have attained mixed results in achieving the goal of improving the conditions of women in our country. It is proud to know that women are entrepreneurs, and successful in many fields and the Women's Reservation Bill being passed by the Government recently. Still it is a sad thing that many villages in our country are striving for their day to day life and toil to make both ends meet. The initiative taken by Mohammad Yunus, Founder of Grameen Bank and the receiver of the Nobel Prize paved the way to encourage women to start up their small business and to fetch their income in order to run their families. He constructed the Self Help Group Model which made the women uplift their families. Replicating the model with suitable modification State and Central government has introduced different interventions for women development.

Statement of the Problem

In general, there are two approaches largely used to examine the empowerment perspective of women. They are:

1. Empowerment resulting in increasing economic status through economic interventions (employment, income generation and access to credit *etc.,*)
2. Integrated rural development programmes (economic status component along with education, literacy, the provision of basic facilities).

Economic wellbeing is not only the single most component considered as individual criteria for measurement of empowerment level of women, which may includes social, psychological, and political, knowledge, creative and entrepreneurial components also. Schemes focused exclusively on women either received reduced allocations or were not implemented, as seen from the revised estimates for 2013-14, almost 87 per cent of 2014-15 budget of MWCP was allocated for Integrated Child development Scheme, leaving only 5 per cent for schemes exclusively meant for women (Hindu dated 25/2/2015). 11th and 12th plan will also commit to continuously focusing on inclusive growth both on development and empowerment perspectives.

Keeping the above said backdrop, a study was conducted to examine empowerment perspective of women as result of enterprising through government sponsored intervention. The specific objective of the study is

- ☆ To examine the empowerment dynamics of selected women SHG.

Methodology

Three SHGs functioning under Thazhavapanchayat, Kollam district of Kerala, Kudumbashree mission, who have engaged in the production of food supplements, were selected purposefully. Sample sizes of 55 active members were selected from three different SHGs as given below.

Sl.No.	*Name of SHGs selected*	*No.of active members (n=55)*	*Enterprising Activities*
1	Ten Star Activity Group, Thazhavapanchayat	19	Production of Health Mix
2	Star Activity Group, Thazhavapanchayat	18	Production of Health Mix
3	Adithya Group, Thazhavapanchayat,	18	Light refreshment food items
	Total	**55**	

The study employed structured interview schedule, focussed group discussion and observation as tools for primary data collection. Secondary data were collected from Grama Panchayath, Kudumbashree mission web site, books related to SHGs and women and also from the institution itself.

The empowerment dynamics was studied by considering the major components (Knowledge Empowerment, Psychological Empowerment, Social Empowerment, Economic Empowerment and Political Empowerment and Entrepreneurial Empowerment selected from the jury opinion. All the major components were measured with the help of identified subcomponents. Each subcomponents listed were measured against the identified items, collected through the process of review of relevant literature, focused discussion with the officials, experts and peer groups. One score was given against each item and thus maximum and minimum score would be 5 and 1, respectively. The obtained score was then categorized as highly empowered (4-5), empowered (3-4), moderately empowered (2-3), little empowered (1-2) and very little empowered (0-1) as the methodology developed and used by Sanyal (2009), Sendilkumar (2011) and (2015). Accordingly responses were collected from the respondents for the two occasions such as before and after joining of membership to the SHG as result of instituional intervention of Kudumbashree mission. The obtained data were analysed using, mean, index and t test for meaningful interpretation. The Empowerment dynamics index was calculated by using the following formula.

$$\mathbf{EDI} = \frac{\text{KEI+PsyEI+SEI+EEI+PEI+ EntEI}}{6}$$

where,

EDI = Empowerment Dynamic Index
KEI = Knowledge empowerment index
PsyEI = Psychological empowerment index
SEI = Sociological empowerment index
EEI = Economic empowerment index
PEI = Political empowerment index
EntEI = Entrepreneurial empowerment index

Findings and Discussion

Knowledge Empowerment

The knowledge empowerment was analyzed in terms of awareness of the information, knowledge and skills possessed by the respondents before and after joining the SHG programme. The knowledge empowerment was conceptualized that knowledge possessed by the respondents on source of raw materials, balanced nutrition, food certification, Good Manufacturing Practices(GMP), food standards and requirements, various women development programmes and schemes.

Table 22.1: Knowledge Empowerment by the Respondents Before and After Joining the SHG (n=55)

Sl.No.	*Knowledge Empowerment Variables*	*Mean Score*		*T Test*
		Before	*After*	
K1.	Knowledge on source of raw materials	1.7273	3.7091	1.79794
K2.	Knowledge on balanced nutrition	1.4727	3.8545	2.17054
K3.	Knowledge on food certification	1.5636	3.7091	1.89912
K4.	Knowledge on (GMP) Good Manufacturing Practices	1.5091	3.7636	1.98312
K5.	Knowledge on various women development programmes and schemes	1.6182	3.7818	1.91541
K6.	Knowledge of different food standards and requirements	1.5636	3.6182	1.73207
K7.	Use of machineries and equipments for food products manufacturing	1.4909	3.6000	1.85596
K8.	Knowledge of value addition on food products	1.5636	3.7091	1.85851
K9.	Knowledge on food adulteration and other existing malpractices	1.7455	3.5091	1.46043
K10.	Knowledge related to availing of banking services	1.3636	2.4727	0.85596
K11.	Knowledge on account keeping	1.2727	2.4000	0.88855
	Total mean Score	**16.8908**	**38.1272**	
	Overall mean Score	**1.53**	**3.46**	

Source: Primary Data : t value (1 per cent significance) = 3.24, t value (5 per cent significance) = 2.66.

Knowledge on balanced nutrition by the respondents have been increased from 1.4, 3.8, and knowledge on source of raw materials was increased from 1.7 to 3.7, respectively after joining the group. Knowledge on Good Manufacturing Practices (GMP) has been drastically increased from 1.5 to 3.7. All the respondents (100 per cent) have answered positively when asked questions about the use of machinery and equipments in food products preparation especially after joining the group. It is also evident from the increase on knowledge dimension on the (K7) use of machinery and equipments for food products manufacturing 1.4 to 3.6,(K9) knowledge on food adulteration and other existing malpractices (1.5 to 3.5), (K10) knowledge of availing banking services(1.3 to 2.6),(K11) knowledge on account keeping(1.2 to 2.4). The mean scores obtained revealed that all dimensions were increased after joining to the SHG programme. The difference in the mean score for knowledge components *viz.*, K1, K2, K3 and K4 were relatively more. Hence it may conclude that women SHGs showed improvement on the knowledge dimension; however it is not statistically significant.

Psychological Empowerment

The psychological empowerment can be assessed in terms of change in courage, decision making quality, risk taking ability *etc.* of the respondents.

Table 22.2: Psychological Empowerment of the Respondents Before and After Joining the SHG (n=55)

Sl.No	*Psychological Empowerment Variables*	*Mean Score*		*T test*
		Before	*After*	
Ps1.	Self confidence	1.8727	4.0727	26.289**
Ps2.	Decision making quality	1.5818	4.0000	23.525**
Ps3.	Risk taking ability	1.6545	3.8545	23.160**
Ps4.	Courage	1.8364	4.2364	20.377**
Ps5.	Motivation in enterprising activities	1.5455	3.8545	26.986**
Ps6.	Positive attitude	2.1818	4.2364	17.521**
Ps7.	Self esteem	1.6727	3.2909	12.866**
	Total mean Score	**12.3454**	**27.5454**	
	Overall mean Score	**1.7636**	**3.9350**	

Source: Primary Data t value (1 per cent significance) = 3.24, t value (5 per cent significance) = 2.66.

Table 22.2 reveals that, there has been considerable improvement in the psychological attributes of the respondents. The confidence level and courage of the respondents has been increased considerably after joining the SHG from the mean score of 1.8 to 4.0 and 1.8 to 4.2 respectively. Remarkable improvement seen in the motivation towards enterprising activities (1.5 to 3.8).This was mainly due to various encouraging policies of the State Government from time to time and involvement of other related agencies in the development of women SHGs. The risk taking ability of the members has been increased after joining the SHG programme. With regard to feeling of positive attitude and self-esteem and decision making ability, there has been an outstanding improvement noticed. The feeling of positive attitude and self-esteem has been increased from 2.1 and 1.6 to 4.2 and 3.2 respectively. The t-test showed a significant difference in the mean score on Ps1, Ps3. Ps4., Ps5. and Ps7. This significant change might be due to the positive orientation with women gender as a part of women empowerment programme of Kudumbasree mission.

Social Empowerment

The empowerment in social activities and its upshot by the respondents were studied in terms of free to work with group members, participation in group activities, involvement in the decision making, participation in gramasabha meetings, free interaction with family members and outsiders, team spirit, leadership quality *etc.* The results obtained are depicted in the Table 22.3.

Table 22.3: Social Empowerment of the Respondents Before and After Joining the SHG Programme (n=55)

Sl.No.	*Social Empowerment Variables*	*Mean Score*		*T Test*
		Before	*After*	
S1.	Free to work with group members	1.8909	4.0364	1.96965
S2.	Participation in the group activities	1.5273	4.1091	2.32859
S3.	Involvement in decision making process	1.7818	4.0727	2.04318
S4.	Attendance in the Gramasabha meetings	1.4364	3.4182	1.63091
S5.	Free interaction with family members and outsiders	1.8909	3.9636	1.76557
S6.	Team spirit and orientation	1.8727	4.1636	2.01232
S7.	Leadership quality	1.7091	4.2182	2.26577
S8.	Group consensus to solve problem	1.7091	3.7273	1.78267
S9.	Developing institutional contact	1.4909	3.2000	1.39835
S10.	Linkage with development departments	1.2727	2.4182	0.86811
S11.	Developing skill to solve conflict	1.7455	3.5273	1.50314
	Total mean Score	**18.3273**	**40.8546**	
	Overall mean Score	**1.6661**	**3.7140**	

Source: Primary Data t value (1 per cent significance) = 3.24 t value (5 per cent significance) = 2.66.

Regarding contact with institutions (S9) and linkage with development departments (S10) by the respondents, remarkable improvement was noticed on the mean score obtained by the members 1.4 to 3.2 and 1.2 to 2.4 respectively. Freeness to work with group members (S1), there was an increase in mean score (1.8 to 4.0), noticed as an indicative improvement as result of being the members of group. In the case of participation in group activities (S2) and involvement in decision making (S3), considerable improvement noticed in the mean score (1.5, 4.1: 1.7, 4.0) observed especially after joining the SHG. From the Table 22.3 it can be seen that team spirit and leadership quality of the respondents were improved and the other components such as free interactions of members of the group with other persons or friends, neighbors *etc.* and attendance to Gramasabha meetings were remains more or less same after joining the group. The t-test showed a difference in the mean score on the focused components; however it is not statistically significant.

Economic Empowerment

The level of economic status together with the social factors determines the position of the persons on the hierarchy. The economic status of the women members in the past and present was studied based on the selected parameters like increase in income, savings habit, investments, financial management skill, extent of dependency on money lenders, use of credit facility in purchasing inputs *etc.*

Table 22.4: Economic Empowerment of the Respondents Before and After Joining the SHG Programme (n=55)

Sl.No.	*Economic Empowerment Variables*	*Mean Score*		*T test*
		Before	*After*	
E1.	Increased income due to group activities	1.6727	3.8727	22.341**
E2.	Availing personal loans and other banking services	1.5273	3.6909	14.794**
E3	Saving money	1.8364	3.9636	17.456**
E4.	Investments in expansion/new venture/assets creation	1.3636	3.96361	17.854**
E5.	Financial management skill	1.4545	3.3273	18.908**
E6.	Improvement in the extent of dependency on money lenders	1.3455	3.7273	12.473**
E7.	Repayment of loans	1.3455	3.0364	12.342**
E8.	Setting down old loans and availing fresh loan for new venture	1.4182	3.0364	12.116**
E9.	Use of credit card	1.2364	2.3091	9.783**
E10.	Use of credit facility in input purchasing	1.2727	2.3091	9.423**
	Total mean Score	**14.4728**	**33.23641**	
	Overall mean Score	**1.4472**	**3.3236**	

Source: Primary Data t value (1 per cent significance) = 3.24 t value (5 per cent significance) = 2.66.

The mean score for the income of the respondents (E1) has been increased from 1.6 to 3.8 after joining the SHG, which might due to the increase in the business turnover. Regarding improvement in the extent of dependency on money lenders, respondents have gained an improved mean score (3.7) after joining the group, *i.e.* the dependency rate of the members was reduced to certain extent. The inputs were purchased from different areas and much care was given for the quality product at reasonable price being the child food products. With respect to availing of loans, they have empowered considerably, because the PACS and such other Commercial Banks (DhanalekshmiBank, Canara bank) are providing financial assistance to the SHGs and thus mean score obtained by the respondents was increased to 3.0 from 1.4. The major sources of savings for the members are income obtained from the enterprising activities of concerned SHGs, and thereby saving money and investments in expansion/new venture/assets creation (1.3 to 3.9). In the case of repayment of loans (E7), the score obtained (3.0) indicates healthy trend. This is mainly due to fact that SHG helped the members to increase their standard of living and financial position to a considerable level and the financial management skill of the women participants increased from 1.4 to 3.2. Slight empowerement was noticed on use of credit card and availing credit facility on purchase of inputs. The t-test showed significant difference in the mean score obtained on all the subcomponents taken for study on economic empowerment dimension.

Political Empowerment

The political empowerment of the members studied in terms of variables like membership in social organization, position assumed in the political party, freedoms of expressing ideas in politics and conflict management. The relevant data was collected and furnished in the Table 22.5.

Table 22.5: Political Empowerment of Respondents Before and After Joining the SHG Programme (n=55)

Sl.No.	*Political Empowerment Variables*	*Mean Score*		*T Test*
		Before	*After*	
P1.	Exercising political affiliation for business order	1.3273	2.3636	8.522**
P2.	Use of political goodwill for continued business	1.3273	2.3455	7.225**
P3.	Members in the social organizations	1.4000	2.7818	10.547**
P4.	Holding position in the political parties	1.1636	2.0545	6.100**
P5.	Freedom of expressing ideas in politics	1.3636	2.5091	8.280**
P6.	Conflict management	2.6727	4.2545	12.524**
	Total mean Score	**9.2545**	**16.309**	
	Overall mean Score	**1.5424**	**2.7181**	

Source: Primary Data t value (1 per cent significance) = 3.24 t value (5 per cent significance) = 2.66.

Table 22.5 reveals that the mean score obtained by the respondents in political empowerment components, before and after the SHG programme is seems to be more or less same. A few respondents were held membership in political parties. With respect to conflict management, the average meanscore obtained by the respondents increased to 4.2. In all other remaining variables, such as membership in social organization, position in the political parties and freedom of expressing ideas in politics, a slight improvement have recorded. The t-test showed a significant difference in the mean obtained score on conflict management. The political association of members helped them to get continued order for the supply of food supplements and products irrespective of the party.

Creativity Empowerment

The factors such as generation of new ideas, orientation towards crisis management were studied in order to analyze the creativity aspect of empowerment component.

Table 22.6: Creativity Empowerment Components of the Respondents Before and After Joining the SHG Programme (n=55)

Sl.No.	Creativity Empowerment Variables	Mean Score		T Test
		Before	After	
C1.	Generate novel ideas for launching products	1.5818	2.8909	11.933**
C2.	Improved level of marketing of products	1.2000	2.4909	9.290**
C3.	Innovative ideas on product development and value addition	1.2545	2.2182	9.938**
	Total mean Score	4.0363	**7.6**	
	Overall mean Score	1.34	**2.53**	

Source: Primary Data t value (1 per cent significance) = 3.24 t value (5 per cent significance) = 2.66.

Table 22.6 reveals that mean scores obtained by the respondents have changed considerably after joining the SHG. With respect to generation of novel ideas for launching products, the average score increased from 1.5 to 2.8, the change was mainly because of the pro active mind set of women members. The efficiency of members in marketing their products was found to be increased (1.2 to 2.4). Similarly, the efficiency for making innovative ideas for product development and value addition had shown increase from 1.2 to 2.2. This was evident from the regular supply order given by the health and nutrition department for the nutritional mixture. The t test showed a significant difference in the mean scores of the empowerment variables on considerable rate.

Entrepreneurial Empowerment

The entrepreneurial empowerment dimension of the members was studied in terms of variables like self-confidence, self-esteem, self-belief, strong sense of purpose, business leadership, integrity, positive attitude, enterprising activities *etc.*, with respect to before and after joining the SHG programmes and furnished in the Table 22.7.

With regard to self-belief integrity and positive attitude, there has been outstanding improvement noticed. After joining the SHG, the mean score obtained was increased from 2.0 to 4.0, 2.2 to 4.1 and 1.8 to 4.2 respectively. Regarding strong sense of purpose and business leadership there was an increase in mean score (1.7 to 3.6 and 1.6 to 3.8), noticed as an improvement after joining the group. Enterprising activities, getting new order than regular, expansion of production capacity, opening outlets, sourcing inputs for production, considerable improvement in the mean score (1.6 to 3.9, 1.2 to 3.2, 1.4 to 3.1 and 1.4 to 3.0) was noticed after joining the SHG. From the Table 22.7 balancing inputs and outputs, business orientation, evolving alternative products, retailing venture were also improved after joining the group. The t-test showed that significant difference in the mean score on the components like En3, En6, En7, En4, En5 and En8 etc.

Table 22.7: Entrepreneurial Empowerment Components of the Respondents Before and After Joining the SHG Programme (n=55)

Sl.No.	*Entrepreneurial Empowerment Variables*	*Mean Score*		*T test*
		Before	*After*	
En1.	Self-confidence	2.0000	3.8727	23.983**
En2.	Self-esteem	1.6545	3.8182	15.824**
En3.	Self-belief	2.0182	4.0364	19.760**
En4.	Strong sense of purpose	1.7273	3.6364	14.932**
En5.	Business Leadership	1.6000	3.8909	24.799**
En6.	Integrity	2.2182	4.1455	18.090**
En7.	Positive attitude	1.8727	4.2182	21.176**
En8.	Enterprising activities	1.6727	3.9091	22.279**
En9.	Getting new order than regular	1.2727	3.2182	13.167**
En10.	Expansion of production capacity	1.4727	3.1455	12.878**
En11.	Opening outlets	1.4364	3.0727	10.998**
En12.	Sourcing inputs for production	2.8545	3.2364	12.866**
En13.	Balancing inputs and outputs	1.4545	3.6545	17.569**
En14.	Business orientation	1.4727	2.6000	19.904**
En15.	Evolving alternative products	1.3636	2.3455	10.752**
En16.	Retailing venture	1.3455	3.8727	9.4000**
	Total mean Score	**27.4362**	**56.6729**	
	Overall mean Score	**1.7147**	**3.5420**	

Source: Primary Data t value (1 per cent significance) = 3.24 t value (5 per cent significance) = 2.66.

Empowerment Dynamics Index (EDI)

All the identified major empowerment components were considered for working out the index.

Table 22.8: Empowerment Dynamics Index of Members in the SHGs

Component	*Index*	
	Before	*After*
Knowledge Empowerment	0.30	0.49
Psychological Empowerment	0.35	0.78
Social Empowerment	0.33	0.74
Economic Empowerment	0.28	0.65
Political Empowerment	0.30	0.54
Creative Empowerment	0.26	0.50
Entrepreneurial Empowerment	0.32	0.69
Empowerment Dynamics	0.30	**0.62**

Table 22.8 reveals that there was only 32 per cent increase in the empowerment dynamics index due to cumulative increase in all the components of empowerment. The increase in EDI was reported for the psychological empowerment component (43 per cent) followed by social empowerment (41 per cent), economic and entrepreneurial empowerment (37 per cent), political and creative ability of the women participants are equally empowered (24 per cent), and finally knowledge empowerment carries the lower empowerment level (*i.e.* 8 per cent). Hence it is concluded that SHG programme has contributed significantly on the empowerment dynamics of women participants who engaged in the production and marketing of food items as a promotional part of Kudumbashree mission, Kerala.

Conclusion

The women SHG programme could empower the rural women in all aspects of life to a great extent. Effective changes occurred in the empowerment of the beneficiaries after joining the group. The results of the study indicated that participation in the SHG programme empowered the women members to get knowledge, psychological abilities, and more involvement in social activities, increased savings habit, better entrepreneurial abilities, freedom from exploitation and avail more political advantages. Members get an opportunity to attend various classes and interact with the authorities to discuss about their problems and finding their own solution. In short, it is conclude that women development programme is a strong support to the rural women and which will make them empowered in all walk of their lives.

References

Sanyal.E. 2009. Empowerment dynamics of GALASA farmers.BSc(Hons) C and B Project, College of Cooperation, Banking and management, Kerala Agricultural University, Thrissur.

Sendilkumar. R. 2011. Empowerment of farmers through GALASA programme – A journey for sustainable agriculture development, Special Issue, Indian Research Journal of Extension Education, Society of Extension Education., Agra. Vol: 12(3): 92-96.

Sendilkumar. R. 2015. Farmers driven value chain of Kadali banana: A gadget for women empowerment. Journal of Krishi Vigyan, Special Issue on Women Empowerment, Society of Krishi Vigyan, Punjab. pp. 44-49.

Chapter 23

Adoption Dynamics of Farmers Field School Beneficiaries in Eco-friendly Technologies for Sustainable Rice Farming

G. Naveenkumar[1] and R. Sendilkumar[2]

[1]*Research Scholar (Agrl. Extension), Department of Agricultural Extension, College of Horticulture, Kerala Agricultural University, Thrissur*
[2]*Professor (Agrl. Extension), Kerala Agricultural University, Thrissur*

ABSTRACT

Rice is the staple food more than 65 per cent of the people of India and cultivated under diverse climatic conditions. To meet the increasing needs of population with limitations on a sustainable basis, the efforts needed are immense and multidimensional. The current thrust is on eco-friendly technologies, whose objective is to exploit only renewable resources to control pollution to tolerable levels and to recycle wastes for future needs. Considering this issue, study was carried out to analyse the adoption of eight identified eco-friendly technologies in rice and constraints encountered by the 120 selected members of Farmers Field School, KVK Palakkad, representing different blocks. Farmers were interviewed and data collected through standardised tool regarding adoption and constraints of eco-friendly technologies of rice. Low marketability and inadequate price to the produce are reported as prime constraints. Complexity in technology, non-availability of inputs, lack of knowledge and skill for determining ETL, low practicability, lack of community participation and adulteration of inputs were also expressed as constraints in adoption of eco-friendly technologies. Strategies to improve the adoption and alleviate the constraints of eco-friendly technologies in rice were suggested.

Keywords: *Adoption, Eco-friendly farm technologies in rice, Farmers field school, Constraints.*

Introduction

The modern agriculture has been successful in meeting the increased food needs of alarmingly growing population, but the problems associated with them are high cost of inorganic or chemical fertilizers, plant protection chemicals, stagnated yield levels and degradation of natural ecosystems. In order to mitigate the health hazards and to bring out natural balance and protection of ecosystem, promotion of eco-friendly technologies through proper skill development by farmers is necessary which can be achieved through promotion of FFS, in which farmers had scope of observing natural ecosystems and their role in control of pests and diseases and conservation of environment.

The advent of chemical intensive farming and its prevalence in Kerala for the past 50 years have resulted in the near stagnant levels of productivity of many of these economically important crops such as coconut, cashew, pepper, coffee, tea, cardamom and arecanut (Sasidharan and Kumar, 2012). The State has attempted to address these issues through innovative scheme such as "Sustainable Development of Rice Based Farming System" and "Macro Management in Agriculture – Rice Development Programme" and given priority to in the annual plans 2010-11, 2011-12. These two schemes targeted to promote rice cultivation through group farming system enabling farmers to adopt improved production technology and scientific package of cultivation suited to each agro-climatic condition and promoting eco-friendly method of pest management like use of bio-pesticides, releasing of predators and parasites and fungal/bacterial pathogens to control pests. There are various factors influencing stakeholders of rice farming in adoption of eco-friendly cultivation practices. Hence, it is very important to understand the adoption dynamics of farmers regarding the eco-friendly technologies. With the above considerations, study entitled was conceived with the objective to study the adoption dynamics of farmer field school (FFS) respondents with respect to eco-friendly farm technologies in rice.

Materials and Methods

The study was carried out in Palakkad district of Kerala State during 2014-15 to analyse the adoption of eco-friendly technologies among the FFS farmers. Rogers's defined adoption as a decision to use and implement a new idea. A total of eight eco-friendly technologies disseminated through FFS were taken. The responses elicited from the respondents were quantified as adoption and non-adoption of the recommended practices. A score of one for adoption, and zero for non-adoption was given. The maximum score that respondents could obtain was 8 and minimum was zero. Depending upon total score obtained by each of the respondent, they were grouped into three categories 'low', 'medium' and 'high' adopter category by using mean and standard deviation (SD) as a measure of check and expressed as a below, and the frequency for adoption of each technology calculated and expressed in percentage.

Eco-friendly Technologies

- ☆ ***Pseudomanas fluorescens*** is common non-pathogenic bacterial antagonist that colonises in soil, water and on plant surfaces. It produces a soluble greenish fluorescent pigment. *P. fluorescens* suppress plant diseases by protecting the seeds and roots from fungal infections. This bacterium is mass-produced using fermentation technology and used for the control of paddy blast and sheath blight and applied as seed treatment(10g/kg of seed), seedling root dip(250g/750 ml),foliar spray(30-45days after transplanting) and soil application.
- ☆ ***Trichoderma viride*** is an antagonistic fungal organism. This biocontrol agent when applied along with seed, colonizes the seed and multiplies on the surface of the seed and kills not only the pathogens present on the surface of the seed but also gives protection against soil-borne pathogens until life time of crop by action of mycoparasitism. It is used against pathogenic fungi *Fusarium, Phytopthara* and *Scelerotia*. Drench the soil near stem region with 10g *Trichoderma* powder mixed in a liter of water for control of Wilt diseases.
- ☆ ***Trichogramma* spp**. belongs to the category of egg parasitoid of biological control agents effective against stem borers of paddy. It is dark coloured tiny wasps and the female wasp lays 20-40 eggs into the host's eggs. The entire cycle is completed within 8-12 days. The tiny adult wasps search for the host (pest) eggs in the field and lay their eggs into the eggs of the pests. The parasitised host's eggs turn uniformly black in 3-4 days. The *Trichogramma* eggs on hatching, feed the embryonic contents of host's egg, completes its development and adult comes out of the host egg by chewing a circular hole. A single *Trichogramma,* while multiplying itself, can thus destroy over 100 eggs of the pest. It attacks the pest at the egg stage itself and hence damage done by larvae is avoided. The recommendation is 5cc/ha.
- ☆ **Pheromones** are natural substances that are produced by special glands in the abdomen of insects and it attracts the opposite gender of the same species. In case of pheromone traps, the lure slowly releases synthetic attractants that helps in detection of a single species of insect. Change of lures should be made at 2-3 week interval (regular interval). Install pheromone traps @ 8/acre for monitoring adult moth activity(Yellow stem borer).
- ☆ **Neem based pesticides**. The most significant liminoids found in neem with proven ability to block insect growth are: azadirachtin, salanin, meliantriol and nimbin. Azadirachtin is currently considered as neem's main agent for controlling insects. It will repel or reduce the feeding of many species of pest insects as well as some nematodes.

- ☆ **Light traps** are mainly used for attracting moths and other night flying insects which are attracted towards the light. It is set in such a way that it encouraged the insect to enter into trap. The lamp (bright white/bluish light) positioned beside a wall or white sheet hang. Set up light traps 1 trap/acre 15 cm above the crop canopy for monitoring and mass trapping insects. Light traps with exit option for natural enemies of smaller size should be installed and operate around the dusk time (6 pm to 10 pm).
- ☆ ***Beauveria bassiana*** is a entomopathogenic fungus that grows naturally in soils and acts as a parasite on various arthropod species, causing white muscardine disease. It is used to control termites, thrips, whiteflies, aphids and lepidopteran pests. As an insecticide, the spores are sprayed on affected crops as an emulsified suspension or wettable powder (20g/lt).
- ☆ **Plant growth promoting rhizobacteria (PGPR)** are the soil bacteria inhabiting around/on the root surface and are directly or indirectly involved in promoting plant growth and development via production and secretion of various regulatory chemicals in the vicinity of rhizosphere. It facilitate the plant growth directly by either assisting in resource acquisition (nitrogen, phosphorus and essential minerals) or modulating plant hormone levels, or indirectly by decreasing the inhibitory effects of various pathogens on plant growth and development in the forms of bio control agents.

Results and Discussion

The adoption level of the eco-friendly farm technologies in rice cultivation by the FFS participants were collected and presented in the Table 23.1 for discussion.

Application of *Pseudomonas*

It is noticed from the Table 23.1 that majority of the respondents (82.00 per cent) had adopted the application of *Pseudomonas fluorescens* and 18 per cent were reported as non-adopters. Higher adoption rate of *Pseudomonas fluorescens* might be due to its attributes like compatibility with the prevailing system and its trialability and also high knowledge index (70.83) of the farmers for this particular technology.

Application of *Trichoderma*

76 per cent of the respondents had adopted the application of *Trichoderma*, whereas remaining 24 per cent of the respondents were non- adopters. The attributes of the technology viz; relative advantage (62.00 per cent) and compatibility (58.00 per cent) and higher knowledge level (67.50 per cent) of the respondents might have promoted the adoption of *Trichoderma* application.

Use of Neem Pesticides and Bio-pesticides

In case of bio pesticides, 72 per cent of the respondents were found to be adopters.It is obvious to note that the bio-control techniques are the innovative practices and require proper scientific knowledge about their use. The possible reasons might be medium to high level of innovativeness, high level extension participation, mass media utilization and knowledge level of the respondents (64.33 per cent) and the attributes of technology viz; relative advantage (82 per cent), compatibility (86 per cent) and observability (83 per cent) might have encouraged the adoption percentage.

Use of Pheromone Traps and Light Traps

Pheromone traps and light traps were adopted by the 72 and 49 per cent of the respondents respectively. The probable reason for adoption might be the FFS participants had opportunities to understand growth and production pattern of the crops and also important aspects like agro ecosystem analysis (AESA). The lessons learnt in AESA might have helped the farmers to know the importance of natural enemies and ETL levels of different pests based on which they had preferred to go for need based spraying.

Table 23.1: Distribution of Respondents on the Basis of Adoption of Individual Eco-friendly Technologies (N =100)

Sl.No.	*Technology*	*Adopters*	*Non-adopters*
		Percentage	*Percentage*
1	Application of *Pseudomonas*	82.00	18.00
2	Application of *Trichoderma*	76.00	24.00
3	Use of *Trichogramma* cards	76.00	24.00
4	Use of pheromone traps	72.00	28.00
5	Use on neem based pesticides and bio pesticides	72.00	28.00
6	Use of light traps	49.00	51.00
7	Application of *Beauveria*	31.00	69.00
8	Application of PGPR	29.00	71.00

Source: Primary data.

Application of PGPR

Majority of the respondents (72 per cent) were found to be in non-adopter category regarding PGPR and only 28 per cent were found to be in adopter category. The possible reason for non-adoption might be lack of adequate knowledge, non-availability and high cost of PGPR along with perceived complexity and low compatibility of the technology.

Application of *Beauveria*

With regard to application of *Beauveria,* 31 per cent had fully adopted the technology, and 69 per cent not adopted it. This might be due to the fact that, non-adopters might have lacked technical skill in handling and using of *Beauveria,* its complexity nature and low compatibility had led to lower adoption. Moreover, the farmers might not have convinced about this practice due to its slow impact on control of pests and intangible nature.

Overall Adoption Level of the Respondents about Eco-friendly Farm Technologies

The distribution of data in Table 23.2 reveals that nearly 70 per cent of the respondents belonged to medium level adoption category, whereas 26 per cent and 4 per cent of FFS respondents belonged to high and low adoption categories respectively.

Table 23.2

Sl.No.	*Category*	*Percentage*
1	Low (Less than 3.00)	4.00
2	Medium (4.00-6.00)	70.00
3	High (Greater than 6.00)	26.00
	Mean - 4.62	**SD- 0.73**

Probable reason for the respondents to be in medium adoption category might be due to the medium to high knowledge possessed by majority (84 per cent) of the respondents. Since knowledge limits the action of individuals, as it is basic pre-requisite for any individuals to think of the pros and cons in making a decision, to either adopt or reject a practice. Other possible reason might be that, majority of the respondents had participated in extension activities like, demonstrations, trainings, group discussions and field days. The results are in line with the findings of Shashidhara (2006).

Constraints Encountered in Adoption of Eco-friendly Technologies

The constraints encountered in adoption of eco-friendly farm technologies furnished in the Table 23.3 and it reveals that 'low marketability and inadequate price to the produce' was perceived as major constraint and ranked first with majority of the respondents in FFS (85.00), followed by complexity in technology (82.00), non-availability of inputs (78.00), lack of knowledge and skill for determining ETL (76.00), low practicability (68.00), lack of community participation (54.00), lack of skill (46.00) and adulteration of inputs (44.00).

The major reason for "Low marketability and inadequate price of the product "might be that there was no exclusive marketing channel and premium support price for the eco-friendly cultivated rice, as everything is procured by Civil Supplies Department in the State made the farmers for not to adopt or discontinued technologies.

Use of *Trichogramma* cards, seedling root dip with *Trichoderma* and *Pseudomonas,* foliar spray with *Beauveria,* Pheromone traps and PGPR were the major technologies being not adopted or discontinued by the rice farmers, because of its complexity nature perceived by farmers in adopting those technologies and hence ranked as second most constraints.

"Non-availability of inputs" was perceived as the third important constraint in adoption of eco-friendly technologies with respect to *Trichogramma,* Pheromone traps, Light traps and PGPR.

Table 23.3 Constraints in Adoption of Eco-friendly Technologies (N=100)

Sl.No.	Constraint	Frequency	Rank
1	Low marketability and inadequate price to the produce	85	I
2	Complexity in technology	82	II
3	Non–availability of inputs	78	III
4	Lack of knowledge and skill for determining ETL	76	1V
5	Low practicability	68	V
6	Lack of community participation	54	VI
7	Lack of skill	46	VII
8	Adulteration of inputs	44	VIII

Source: Primary data.

Pheromone traps and Light traps were the major technologies being not adopted or discontinued due to the difficulty in remembering practical approach of ETL levels for each and every pest.

Pheromone traps, seedling root dip with *Trichoderma* and *Pseudomonas*, *Trichogramma*, foliar spray with *Beauveria* and Light traps were the major technologies being not adopted or discontinued by the rice farmers because of low practicability of those technologies at the contingent situation.

"Lack of Community participation" was reported as one of the constraint in adoption of eco-friendly farm technologies in rice cultivation and was ranked sixth by rice farmers, because all these practices should require co-operation and participation from neighboring farmers for achieving good results in compact area. Lack of community participation results slow down the process of diffusion.

Application of *Trichogramma*, Pheromone traps, seed treatment and PGPR requires adequate skill for its effectiveness and better results. Lack of skill in handling those technologies might be the reason poor adoption.

Neem cake, *Pseudomonas* and *Trichoderma* application were the major technologies being not adopted or discontinued by the rice farmers because of lack of good quality inputs with outlets reduces the effectiveness of the technology. "Adulteration of the inputs" was expressed as minor constraint in adoption of eco-friendly farm technologies.

Suggestions to Enhance Adoption of Eco-friendly Technologies

Based on the critical analysis of different constraints in adoption of eco-friendly farm technologies in rice, extension strategy is proposed to overcome the constraints for effective adoption.

- ☆ **Imparting knowledge**: Extension activities like training programmes supplemented with farm literature in local language to be emphasized for the spread of eco-friendly farming technologies.
- ☆ **Imparting skills:** In general application of technologies requires more of skills to take up in an effective way so as to attain better results. Hence,

more number of method demonstrations need to be conducted exclusively to these technologies supplemented with the extension literature.

- ☆ **Creating awareness:** There is a need for efficient utilization of mass media such as Radio, TV, newspaper for creating awareness on the existing technologies and their impact of crop production need to be explained. Hence the priority should be given for creating awareness by utilizing different mass media and reinforced through group and personal contacts.
- ☆ **Input availability:** The technologies like *Trichogramma*, Pheromone traps, light traps, PGPR, *Beauveria* and bio pesticides should be available at right quantity,right quality and right time. Hence such critical inputs need to be made available to the pro poor farmers with subsidy component.
- ☆ **Technology assessment and refinement:** The technologies like Tricho cards, dipping of nursery bundles in bio control agents, Pheromone traps, application of right chemical with right dosage and method, application of PGPR, light traps may require further simplification at the field level. Because of its perceived complexity, labour intensity and low practicability in adoption the farmers have ignored such technologies. Hence these technologies should be further refined in such a way that they are easy to adopt, involves less labour and also have more practicability.
- ☆ **Credit availability**: Irrespective of size of land holdings either own or leased, farmers need to be provided with timely credit for taking up farm operations. This facilitates in adoption of not only eco-friendly technologies, but also the other technologies which are useful for better productivity.
- ☆ **Input quality standards:** The technologies like neem based chemicals; bio control agents, dipping of nursery bundles in solution of *Pseudomonas, Trichoderma* and foliar application of *Beauveri* require strict vigilance and inspection to maintain its quality standards. Hence, necessary steps to be initiated in that dimension and see that farmers should get quality inputs.
- ☆ **Price of the product:** An exclusive price policy had to be formulated for the produce of eco-friendly farm technologies by the government. Promotion of Farmer Producer Companies for the processing and value addition of eco-friendly products had to be done for getting remunerative price to the farmers which enhances adoption for large scale expansion of eco-friendly farming in rice.

Conclusion

Each eco-friendly technology is having its own set of constraints for adoption by the farmers. There is a need to design extension strategies based on the specific constraints faced by the farmers for each eco-friendly technology. Blending of different extension approaches for different groups of eco-friendly technologies will give better results for the committed extension efforts of the extension personnel. Hence, there is an immediate need to promote eco-friendly method of cultivation, focusing more on imparting the principles of FFS during the training programmes and demonstrations, skill development among rural youth and farmers.

References

Naveenkumar.G 2015. Impact of eco-friendly technologies in Rice dissimintated through Farmers field School. MSc (Ag) thesis, Kerala Agricultural University, Thrissur.

Shashidhara.2006 A study on management of eco-friendly practices by vegetable growers of north Karnataka. MSc (Ag) thesis, University of Agricultural Sciences, Dharward. 90 p.

Sasidharan D and Kumar. P, 2012. Trend analysis of Paddy cultivation in Kerala. J. Ext. Educ.15: 61-62.

Index

Index

E

F

G

H

I

K

K

M

N

www.ingramcontent.com/pod-product-compliance
Ingram Content Group UK Ltd.
Pitfield, Milton Keynes, MK11 3LW, UK
UKHW021531300726
14060UKWH00011B/343

9 789388 173872